南京大学建筑与城市规划学院 建筑系
教学年鉴 2011—2012
THE YEAR BOOK OF ARCHITECTURE PROGRAM
2011–2012, Volume 12
SCHOOL OF ARCHITECTURE AND URBAN PLANNING
NANJING UNIVERSITY

华晓宁 刘铨 王丹丹 编　Editors: HUA Xiaoning LIU Quan WANG Dandan
东南大学出版社·南京　Southeast University Press, Nanjing

图书在版编目（CIP）数据

南京大学建筑与城市规划学院建筑系教学年鉴. 2011
~2012 / 华晓宁，刘铨，王丹丹编. -- 南京：东南大
学出版社，2012.12
　　ISBN 978-7-5641-3970-4

Ⅰ. ①南… Ⅱ. ①华… ②刘… ③王… Ⅲ. ①建筑学
—教学研究—高等学校—南京市—2011~2012—年鉴②城
市规划—教学研究—高等学校—南京市—2011~2012—年
鉴 Ⅳ. ①TU-42

中国版本图书馆CIP数据核字（2012）第295722号

策　　划：丁沃沃　赵　辰
装帧设计：刘　铨　华晓宁
参与制作：黎建波　袁金燕

出版发行：东南大学出版社
社　　址：南京市四牌楼2号
出 版 人：江建中
网　　址：http://www.seupress.com
邮　　箱：press@seupress.com
邮　　编：210096
经　　销：全国各地新华书店
印　　刷：南京新世纪联盟印务有限公司
开　　本：787mm×1092mm　1/20
印　　张：8
字　　数：480千
版　　次：2012年12月第1版
印　　次：2012年12月第1次印刷
书　　号：ISBN 978-7-5641-3970-4
定　　价：48.00元

本社图书若有印装质量问题，请直接与营销部联系。电话：025-83791830

建筑设计及其理论
Architectural Design and Theory

张 雷 教 授　　　Professor ZHANG Lei
冯金龙 教 授　　　Professor FENG Jinlong
吉国华 教 授　　　Professor JI Guohua
周 凌 副教授　　　Associate Professor ZHOU Ling
傅 筱 副教授　　　Associate Professor FU Xiao
胡友培 讲 师　　　Lecturer HU Youpei

城市设计及其理论
Urban Design and Theory

丁沃沃 教 授　　　Professor DING Wowo
华晓宁 副教授　　　Associate Professor HUA Xiaoning
许 浩 副教授　　　Associate Professor XU Hao
刘 铨 讲 师　　　Lecturer LIU Quan
尹 航 讲 师　　　Lecturer YIN Hang

建筑历史与理论及历史建筑保护
Architectural History and Theory, Protection of Historic Building

赵 辰 教 授　　　Professor ZHAO Chen
王骏阳 教 授　　　Professor WANG Junyang
肖红颜 副教授　　　Associate Professor XIAO Hongyan
胡 恒 副教授　　　Associate Professor HU Heng
冷 天 讲 师　　　Lecturer LENG Tian

建筑技术科学
Building Technology Science

鲍家声 教 授　　　Professor BAO Jiasheng
秦孟昊 教 授　　　Professor QIN Menghao
吴 蔚 副教授　　　Associate Professor WU Wei
郜 志 副教授　　　Associate Professor GAO Zhi
童滋雨 讲 师　　　Lecturer TONG Ziyu

南 京 大 学 建 筑 与 城 市 规 划 学 院 建 筑 系
Department of Architecture
School of Architecture and Urban Planning
Nanjing University
arch@nju.edu.cn　　　http://arch.nju.edu.cn

教学纲要

教学阶段 Phases of Education	本科生培养（学士学位）Undergraduate Program (Bachelor Degree)
	一年级 1st Year / 二年级 2nd Year / 三年级 3rd Year / 四年级 4th Year
教学类型 Types of Education	通识教育 General Education / 专业教育 Professional Training
课程类型 Types of Courses	通识类课程 General Courses / 学科类课程 Disciplinary Courses / 专业类课程 Professional Courses
主干课程 Design Courses	设计基础 Basic Design / 建筑设计基础 Basic of Architectural Design / 建筑设计 Architectural Design
理论课程 Theoretical Courses	专业基础理论 Basic Theory of Architecture / 专业理论 Architectural Theory
技术课程 Technological Courses	
实践课程 Practical Courses	环境认知 Environmental Cognition / 古建筑测绘 Ancient Building Survey and Drawing / 工地实习 Practice of Construction Plant

EDUCATIONAL PROGRAM

研究生培养（硕士学位）Graduate Program (Master Degree)			研究生培养（博士学位） Ph. D. Program
一年级 1st Year	二年级 2nd Year	三年级 3rd Year	

学术研究训练 Academic Research Training

学术研究 Academic Research

建筑设计研究 Research of Architectural Design	毕业设计 Thesis Project	学位论文 Dissertation	学位论文 Dissertation

专业核心理论 Core Theory of Architecture	专业扩展理论 Architectural Theory Extended	专业提升理论 Architectural Theory Upgraded	跨学科理论 Interdisciplinary Theory

建筑构造实验室 Tectonic Lab
建筑物理实验室 Building Physics Lab
数字建筑实验室 CAAD Lab

生产实习 Practice of Profession   生产实习 Practice of Profession

课程安排

	本科一年级 Undergraduate Program 1st Year	本科二年级 Undergraduate Program 2nd Year	本科三年级 Undergraduate Program 3rd Year
设计课程 Design Courses	设计基础 Basic Design	建筑设计基础 Basic Design of Architecture 建筑设计（一） Architectural Design 1	建筑设计（二） Architectural Design 2 建筑设计（三） Architectural Design 3 建筑设计（四） Architectural Design 4 建筑设计（五） Architectural Design 5
专业理论 Architectural Theory	逻辑学 Logic	建筑导论 Introductory Guide to Architecture	建筑设计基础原理 Basic Theory of Architectural Design 居住建筑设计与居住区规划原理 Theory of Housing Design and Residential Planning 城市规划原理 Theory of Urban Planning
建筑技术 Architectural Technology	理论、材料与结构力学 Theoretical, material & structural Statics Visual BASIC程序设计 Visual BASIC Programming	CAAD理论与实践 Theory and Practice of CAAD	建筑技术（一） 结构与构造 Architectural Technology 1: Structure Construction & Execution 建筑技术（二） 建筑物理 Architectural Technology 2: Building Physics 建筑技术（三） 建筑设备 Architectural Technology 3: Building Equipment
历史理论 History Theory	古代汉语 Ancient Chinese	外国建筑史（古代） History of World Architecture (Ancient) 中国建筑史（古代） History of Chinese Architecture (Ancient)	外国建筑史（当代） History of World Architecture (Modern) 中国建筑史（近现代） History of Chinese Architecture (Modern)
实践课程 Practical Courses		古建筑测绘 Ancient Building Survey and Drawing	工地实习 Practice of Construction Plant
通识类课程 General Courses	数学 Mathematics 语文 Chinese 名师导学 Guide to Study by Famed Professors 计算机基础 Basic Computer Science	社会学概论 Introduction of Sociology 社会调查方法 Methods for Social Investigation	
选修课程 Elective Courses		城市道路与交通规划 Planning of Urban Road and Traffic 环境科学概论 Introduction of Environmental Science 人文科学研究方法 Research Method of the Social Science 美学原理 Theory of Aesthetics 管理学 Management 概率论与数理统计 Probability Theory and Mathematical Statistics 国学名著导读 Guide to Masterpieces of Chinese Ancient Civilization	人文地理学 Human Geography 中国城市发展建设史 History of Chinese Urban Development 欧洲近现代文明史 Modern History of European Civilization 中国哲学史 History of Chinese Philosophy 宏观经济学 Macro Economics 管理信息系统 Management Operating System 城市社会学 Urban Sociology

CURRICULUM OUTLINE

本科四年级	研究生一年级	研究生二、三年级
Undergraduate Program 4th Year	Graduate Program 1st Year	Graduate Program 2nd & 3rd Year
建筑设计（六） Architectural Design 6	建筑设计研究（一） Design Studio 1	专业硕士毕业设计 Thesis Project
建筑设计（七） Architectural Design 7	建筑设计研究（二） Design Studio 2	
本科毕业设计 Graduation Project	数字建筑设计 Digital Architecture Design	
	联合教学设计工作坊 International Design Workshop	
城市设计理论 Theory Urban Design	城市形态研究 Study on Urban Morphology	
	现代建筑设计基础理论 Preliminaries in Modern Architectural Design	
	现代建筑设计方法论 Methodology of Modern Architectural Design	
	景观都市主义理论与方法 Theory and Methodology of Landscape Urbanism	
建筑师业务基础知识 Introduction of Architects' Profession	材料与建造 Materials and Construction	
建设工程项目管理 Management of Construction Project	中国建构（木构）文化研究 Studies in Chinese Wooden Tectonic Culture	
	计算机辅助技术 Technology of CAAD	
	GIS基础与运用 Concepts and Application of GIS	
	建筑理论研究 Study of Architectural Theory	
生产实习（一） Practice of Profession 1	生产实习（二） Practice of Profession 2	建筑设计与实践 Architectural Design and Practice
景观规划设计及其理论 Theory of Landscape Planning and Design	建筑史研究 Studies in Architectural History	
东西方园林 Eastern and Western Gardens	建筑节能与可持续发展 Energy Conservation & Sustainable Architecture	
地理信息系统概论 Introduction of GIS	建筑体系整合 Advanced Building System Integration	
欧洲哲学史 History of European Philosophy	规划理论与实践 Theory and Practice of Urban Planning	
微观经济学 Micro Economics	景观规划进展 Development of Landscape Planning	
政治学原理 Theory of Political Science		
社会学定量研究方法 Quantitative Research Methods in Sociology		

目录

教学论文

探索建筑教育的多元化模式	丁沃沃	2
通识教育背景下建筑系本科设计课程设置的探索	周凌　丁沃沃	4
南京大学外国建筑史教学经验谈	胡　恒	8
设计本身节能的健康建筑	秦孟昊	12
建筑技术课程的创新探索	吴　蔚	14
CAAD基础教学新思路	童滋雨　刘　铨	20

常设设计课程

建筑设计基础	刘　铨　冷　天	24
建筑设计（一）　小型公共建筑设计	刘　铨　冷　天	26
建筑设计（二）　小型建筑设计	周　凌　童滋雨　钟华颖	30
建筑设计（三）　中型公共建筑设计	周　凌　童滋雨　钟华颖	34
建筑设计（四）　大型公共建筑设计	华晓宁　胡友培　王丹丹	38
建筑设计（五）　住宅小区规划设计	华晓宁　胡友培　王丹丹	42
建筑设计（六）　城市设计	丁沃沃　尹　航　胡友培	46
建筑设计（七）　高层建筑设计	吉国华　胡友培　尹　航	50
建筑设计研究（一）　基本设计	张　雷　傅　筱	54
建筑设计研究（一）　概念设计	周　凌　张　旭　刘可南	58
建筑设计研究（二）　建构设计	傅　筱　郭屹民	62
建筑设计研究（二）　城市设计	丁沃沃　冯　路	66

毕业设计

本科毕业设计	72
专业硕士毕业设计：城市设计研究	84
专业硕士毕业设计：公共建筑设计	88
专业硕士毕业设计：建构设计研究	92
专业硕士毕业设计：绿色建筑设计	96
专业硕士毕业设计：建筑遗产再生	100

设计工作坊

古建筑测绘	赵　辰　萧红颜	106
中国建构（木构）文化研究	赵　辰　冯金龙	108
数字建筑设计	吉国华	112
影像南京	Marc Boumeester	116
体积编码（南京）	Tom Kvan & Justyna Karakiewicz	118
虚拟餐厅	Doris Fach & Hans Sebastian von Bernuth	120

建筑理论课程	123
城市理论课程	127
历史理论课程	131
建筑技术课程	135

其他

讲座	140
硕士学位论文列表	142

TABLE OF CONTENTS

Articles on Education

Page	Authors	Title
2	DING Wowo	Exploration on the Diversified Mode of Architectural Education
4	ZHOU Ling, DING Wowo	Exploration on the Setup of Undergraduate Design Courses for the Department of Architecture under the General Education Background
8	HU Heng	On Nanjing University's Experience in Foreign Architectural History Teaching
12	QIN Menghao	Design for Green and Healthy Building
14	WU Wei	An Innovation Experiment of Building Technology Teaching in an Architecture School
20	TONG Ziyu, Liu Quan	New Ideas about Basic CAAD Teaching

Regular Design Courses

Page	Authors	Title
24	LIU Quan, LENG Tian	Basic Design of Architecture
26	LIU Quan, LENG Tian	Architectural Design 1: Small Public Building
30	ZHOU Ling, TONG Ziyu, ZHONG Huaying	Architectural Design 2: Small Building
34	ZHOU Ling, TONG Ziyu, ZHONG Huaying	Architectural Design 3: Public Building
38	HUA Xiaoning, HU Youpei, WANG Dandan	Architectural Design 4: Complex Building
42	HUA Xiaoning, HU Youpei, WANG Dandan	Architectural Design 5: Residential Planning
46	DING Wowo, YIN Hang, HU Youpei	Architectural Design 6: Urban Design
50	JI Guohua, HU Youpei, YIN Hang	Architectural Design 7: High-rise Building
54	ZHANG Lei, FU Xiao	Design Studio 1: Basic Design
58	ZHOU Ling, ZHANG Xu, LIU Ke·nan	Design Studio 1: Conceptual Design
62	FU Xiao, GUO Yimin	Design Studio 2: Constructional Design
66	DING Wowo, FENG Lu	Design Studio 2: Urban Design

Graduation and Thesis Projects

Page	Title
72	Graduation Project
84	Thesis Project: Urban Design
88	Thesis Project: Public Building Design
92	Thesis Project: Construction Design
96	Thesis Project: Green Building Design
100	Thesis Project: Building's Regeneration

Design Workshops

Page	Authors	Title
106	ZHAO Chen, XIAO Hongyan	Ancient Building Survey and Drawing
108	ZHAO Chen, FENG Jinlong	Studies in Chinese Wooden Tectonic Culture
112	JI Guohua	Digital Architecture Design
116	Marc Boumeester	Nanjing Cinematic
118	Tom Kvan & Justyna Karakiewicz	Coding Volumetric Nanjing
120	Doris Fach & Hans Sebastian von Bernuth	"Eat out": Imaginations of a Dining Hall in Nanjing

Miscellanea

Page	Title
123	Architectural Theory
127	Urban Theory
131	History Theory
135	Architectural Technology
140	Lectures
142	List of Thesis for Master Degree

教学论文
ARTICLES ON EDUCATION

探索建筑教育的多元化模式 • 丁沃沃

建筑学在西方已有大约500年的历史,是西方大学里的传统学科。上个世纪之初随着西学东进的潮流中国引入了西方建筑学教育,因此对于中国来说,建筑学教育历史远远短于我们辉煌的建筑历史。以西方建筑学为基础的我国建筑教育体系近百年来为国家培养了大量的职业建筑师,尤其是近三十年来新一代的建筑师在我国城市化进程中发挥了巨大的作用。随着社会发展的需要,我国目前不仅需要大量的城市建设者,也需要更多有专业知识的城市经营者和管理者,也更需要高质量的设计师。因此,作为建筑教育工作者应该根据国家的需要,在不同历史阶段对建筑学教育作相应调整。

从社会需求来看,城市化进程的提速和城市建设质量的提高将使国家对建筑师的需求由数量转向质量。从行业发展趋势来看,高端专业人才的培养直接关系到国家各行业整体质量的提升。就建筑学专业学位而言,它的设立意在提高应用型人才的专业能力和综合素质,是在原有建筑学培养计划的基础上加强应用技术的输入、行业规范的教育以及实际操作能力的训练,而不能理解为将原有的建筑学教育降低为以操作性为主的职业教育。建筑学学位的多样化应该对应建筑教育的多样化,因此,探索适合我国经济发展模式、发展阶段,同时和国际接轨的有自身特色的建筑学学位教育模式非常有必要,且意义重大。为此,南京大学建筑学科开始研究建筑学教育的发展趋势和大学多学科基础对建筑学教育的影响,探索多学科交叉背景下建筑学教育的新思路。

在分析了国内外一流大学建筑教育的基础上,我们认为现行的建筑教育存在两个主要问题:a. 专业学位过于重复,即本科五年学生已经获得了可以参加执业资格考试的建筑学学士,在研究生阶段又须三年获得一个建筑学硕士。两个专业学位共耗时8年,仅层次不同,对执业资格考试没有本质的影响。对学生来说学制过长,对学校来说浪费了有限的资源。b.过早强化专业知识。由于本科要拿专业学位,所以只能压缩一般大学知识尽早进入专业教育以满足专业学位的学时数。知识基础不够宽,研究视野不够广,方法不系统和学术不规范等问题,使得学生后续发展空间受到限制。相对于国际一流大学,我们的建筑学学术研究一直在较低水平上徘徊。据此,我们认为应重新整理教育思路,结合国情,建构分类贯通的建筑学人才培养新模式。

为此,南京大学建筑学科首先对自己的培养目标进行定位,在学院全体教师的充分讨论和论证下,根据南京大学的性质、地位和资源,决定了建筑学人才培养的目标:建筑学高端人才。根据国家的需要,结合我国国情和建筑学的学科特色,我们设定了分层次、分类型的人才培养方案。尤其是结合专业教育和学术教育的不同要求,注重三类人才的培养:应用型、复合型和学术型。

1. 高层次应用型人才:随着我国城市化进程的提速和城市建设质量的提高,国家对建筑师的需求将会由数量向质量转变。从发展趋势来看,高端建筑设计专业人才的培养直接关系到国家建筑行业整体质量的提升,国家目前迫切需要建筑学高端应用型人才。

2. 高层次复合型人才:我国正处于城市化发展的快速且关键阶段,新理念、新事物和新行业不断变化,因此具有宽基础的有应变能力的复合型人才对国家的发展与建设做出贡献。就建筑行业而言,建筑的开发行业和管理行业都需要具有建筑学背景的复合型专业人才。为此,美国一流大学在本科一、二年级实行的不分专业的通识教育有借鉴的价值,这种机制提供了后期在高端教育中跨学科培养的可能。

3. 高层次学术型人才:从提高国家竞争力的角度看,高层次学术型人才的培养不仅是提升国家整体实力的需要,也是衡量教育和科学水平的需要。随着我国综合国力的提升,对高层次学术型人才的需求会越来越迫切。高层次人才培养是一个系统工程,并不能一蹴而就。为此,高层次人才的培养必须从基础抓起。只有广阔的学识才能支撑起有高度的学术空间。

在培养目标和参照标准明确之后,南京大学建筑学科在同行的支持和鼓励下,开始了建筑教育模式的全面改革:由建筑教育的单一化(培养职业建筑师)模式改革为多元化模式,即为整个建筑行业的需求培养人才。具体的构架是:本科实行多目标、宽口径的通识教育培养模式,同时为研究生阶段的专业教育打下坚实而宽阔的基础;研究生阶段实行国际通行的课程体系培养高层次专业人才,同时以实习基地的模式结合国情强化学生的操作能力。博士阶段走国际化之路,将研究课题直接和国际接轨。

经过为期6年的探索,初步形成了一套较为完整的体系。在学制的设置方面吸取了国际一流大学建筑学的办学经验,同时结合我国国情和建筑学的学科特色,建构了分层次分类型地建构人才培养方案,即:2(通识)+2(专业)+2或3(研究生)模式。具体地说就是第一个2年设定为建筑学的通识教育阶段,其中包括一年级的大学通识和二年级的建筑学基础通识;第二个2年为建筑学专业教育,强化专业知识;最后的2(或3)年为研究生教育,同时设立了三个培养目标,即专业型教育、复合型教育和学术型教育。我们针对各类人才的需求,对培养模式做了进一步细化,即2+2+2的专业型人才培养、2+2+3的复合型人才培养、2+2+6的学术型人才培养等模式,这样就构成了我们多元化建筑学高层次人才的培养体系。

在课程体系的设置方面,我们概括为:一条主干、四个类别、多项选择。以模块化的课程组合构架出不同的课程体系,实现了以核心课程为主干的开放式、分类型的高层次人才培养模式。

一条主干是指以设计训练为主干,其中包括了各类别的基础设计、建筑设计和城市设计。四个类别的课程分别为:基本知识、课程设计、理论训练、设计实践。多项选择即:各类跨学科选修课、各类设计工作坊、国际合作教学和基地实习。由于各类人才的培养分类在研究生阶段,因此我们主要对研究生课程进行整合、重组,实现课程多样化,做到了精炼必修课程,强化专业核心课程,针对不同类型的人才培养模式设置不同的课程模块,根据不同类型进行选择和组合。

南京大学建筑学科多元化培养模式力求做到四个相结合:前瞻性和可操作性相结合;实践性训练和研究性训练相结合;规范性设计与创意性设计相结合;国际化视野和中国特色相结合。该模式实行6年来,从学生的综合素质来看,该培养模式基本达到了预期的效果。对于南京大学建筑学教育来说,本科的培养仅仅是阶段性成果,研究生阶段的多样化教育是建筑教育的核心。

The architecture in the western countries has a history of about 500 years, and it is one of the traditional disciplines of western universities. With the trend of "Learning from the western countries and introducing the knowledge into the Oriental countries" at the beginning of the last century, China has introduced a western architectural education. So, for China, the history of architectural education is far shorter than our splendid architectural history. Over the past century, China's architectural education system that is based on the western architecture has cultivated a large number of professional architects for the nation; especially in the past three decades, the new generation of architects has played a huge role in the urbanization process of China. With the needs of social development, contemporary China needs not only a large number of city builders, but also more urban operators and managers with expertise, as well as qualified designers. Therefore, educators should make relevant adjustment to the architectural education in different historical stages according to the needs of the country.

From the perspective of social demands, the speed of the process of urbanization and the improvement of the quality of urban construction will make the national demand for architects change from quantity to quality. From the perspective of industrial trends, the training of high-end professionals relates directly to the enhancement of the overall quality of various industries of China. And from the perspective of professional degree of architecture, it's setup intends to improve the professional competence and overall quality of the applied talents, and to strengthen the input of the applied technology, the education of the industry standards and the training of actual operating capacity based on the original architectural training plan, and it cannot be understood as lowering the original architectural education to the vocational education focusing on the operability. The diversification of the degree of architecture should correspond to the diversification of architectural education. Therefore, it is necessary and important to explore the mode of architectural degree education that is suitable for China's economic development model and stage of development, and has its own distinctive feature conforming to international standards. Therefore, the architectural education in Nanjing University has started to

EXPLORATION ON THE DIVERSIFIED MODE OF ARCHITECTURAL EDUCATION • DING Wowo

study the development trend of architectural education and the impact of university multidisciplinary basis on the architecture, and explore the new ideas of architectural education under the background of the crossed multidiscipline.

Based on the analysis of the architectural education in the leading universities at home and abroad, we believe that the current architectural education has two major problems: a. the professional degree is too repetitive, i.e. a five-year undergraduate student has acquired the degree of bachelor of architecture necessary for participating in the licensing examination, and he or she needs another three years to acquire a degree of master of architecture at the graduate level. The two degrees are of different levels and take total of 8 years, having no essential impact on the licensing examination. For the students, the school system is too long; and for the universities, it is a waste of the limited resources. b. premature strengthening on the expertise. Since the undergraduate students have to obtain professional degrees, we can only compress the general academic knowledge teaching as soon as possible and enter the stage of professional education to achieve the required hours for professional degrees. Such problems as the knowledge base is not wide enough, the research vision is not open enough, the method is not systematic enough and the academic is not standardized enough, etc., resulting in the students' subsequent development space being restricted. Compared to world-class universities, our academic research on architecture is hovering at a lower level. Accordingly, we consider refreshing the ideas of education and combining with the national conditions to construct a new classification-through mode of architectural talent training.

To this end, Nanjing University positions the objectives of its architectural training first. Upon a full discussion and argumentation among all the teachers of the school and based on the nature, status and resources, the objectives of architectural talent training are determined as: high-end architectural talent. According to the needs of national development and in combination with China's national conditions and the disciplinary characteristics of architecture, we establish a hierarchical and classified talent training program. Especially in combination with the different requirements of professional education and academic education, we focus on the training of three types of talents: applied type, versatile type, and academic type.

1. High-level applied talents: With the acceleration of the urbanization process and the improvement of the quality of urban construction in China, the focus of the national demand for architects will change from the quantity to quality. From the perspective of development trends, the training of high-end architectural design professionals relates directly to the enhancement of the overall quality of the national construction industry, and China has now an urgent need for high-end applied talents in the field of architecture.

2. High-level versatile talents: China is now at a rapid and critical state of development in urbanization, and new ideas, new things, and new industries are constantly changing, versatile talents with a broad-base and types of capacity will therefore make a great contribution to the development and construction of the country. As for the architectural industry, the architectural development and management industries need the versatile professionals with architecture background. Therefore, the general education, regardless of specialties practiced by the best American universities in undergraduate grade 1 and grade 2, is of reference value. This mechanism provides the possibility of the interdisciplinary training in the late high-end education.

3. High-level academic talents: From the perspective of enhancing national competitiveness, the training of high-level academic talents is not only a need for enhancing overall national strength, but also a need for measuring the educational and scientific levels. With the enhancement of China's comprehensive national strength, the demand for high-level academic talents will become increasingly urgent. The training of high-level talents is a systematic project and cannot be done overnight. Therefore, the training of high-level talents must start from the basics. Only broad knowledge can prop up a high level of academic space.

After clarifying its training objectives and reference standard, the architectural disciplines in Nanjing University initiates the overall reform of architectural education system under the support and encouragement of the peers: changing from the simple (training vocational architects) mode of architectural education to the diversified mode of architectural education, i.e. training talents for the demands of the whole architectural industry. The specific framework: at the undergraduate stage, adopt the training mode of multi-target and wide-caliber general education, and lay a solid and broad foundation for specialized education at the postgraduate stage; at the postgraduate stage, use the internationally accepted curriculum system to train high-level professionals, and strengthen the students' operating ability using the practice base mode in combination with the national conditions; and at the doctoral stage, take an international way to link the research projects to the international standards.

After a six-year exploration, a set of more complete systems was initially formed. The setup of the school system has learned from the experience of the architecture education of the world-class universities and combines China's national conditions with the disciplinary feature of architecture to develop a hierarchical and classified talent training program, i.e.: 2 (General) +2 (professional) +2 or 3 (postgraduate) mode. Specifically, the first two years are for the general education stage of architecture, including the university general education in first year and the general architecture basis in the second year; the second two years are for the professional education of architecture which strengthens expertise; and the final 2 (or 3) years are for graduate education, and three training objectives are set at the same time, i.e. the professional education, versatile education and academic education. For the demands for various types of talents, we further refine the training mode, i.e. the training of 2+2+2 type of professionals, the training of 2+2+3 type of versatile talents and the training of 2+2+6 type of academic talents, forming our diversified training system for high-level talents of architecture.

In setting up the curriculum, we summarize as follows: one backbone, four categories, and multiple options. Develop a different curriculum system with the modular curriculum combination to implement the open and classified training mode for high-level talents that takes the core curriculum as the backbone.

One backbone means taking the architectural design training as a backbone, including the fundamental design, architectural design, and urban design in various categories. The four categories of courses include: basic knowledge, curriculum design, theoretical training, and design practice. Multiple options: the various interdisciplinary elective courses, all kinds of design workshops, international cooperative teaching, and practice at the base. Since the training of various types of talents is classified at the graduate stage, we mainly carry out integration and restructuring of postgraduate courses, thus implementing curriculum diversification. We have realized refining the required courses and strengthening the professional core courses. We set different course modules for different types of talent training mode and make selection and combination according to the different types.

The diversified training mode of architecture in Nanjing University strives to achieve four combinations: the combination of forward-looking and operability, the combination of practical training and research training, the combination of standardized design and creative design and the combination of international vision and Chinese characteristics. Since the 6 years of the implementation of this mode, this training mode has basically achieved the desired effect in the aspect of the students' comprehensive qualities. For the educators of architecture in Nanjing University, the training of undergraduates is just a periodical achievement, while the diversified education at the postgraduate stage is the core of architectural education.

通识教育背景下建筑系本科设计课程设置的探索 • 周凌 丁沃沃

一、背景：培养什么样的人才

目前我国的建筑学教育虽然套用了国际上通行的专业学位教育概念，但是在模式上并没有完全对应，主要存在三大问题：a.建筑学专业学位重复设置，本科和硕士没有明确分工，导致硕士专业学位的学制过长，教育资源浪费。b.就研究型大学而言，由于本科以专业学位为出口，不得不压缩通识课程而过早进入专业训练，使得学生后续发展空间受到限制。该现状直接导致了本学科高层次研究型人才的缺乏，造成建筑学的研究和创新落后于国家高端需求。c.同样由于通识类知识匮乏，学生的学术视野较窄，远远不能满足整体建筑行业对人才能力多样化的需求，更不能满足国家未来发展的新要求。

为解决上述问题，南京大学建筑与城市规划学院在2007年开始进行建筑教育模式的改革和创新，在理念、方法和操作措施方面进行了研究与探索，主要有几方面：

1. 建筑学通识教育和专业教育相结合的新模式，即：4（本科）+ 2/3（研究生）模式。参照国际一流大学的专业定位，将建筑学专业教育的出口放在研究生层面，学生获得建筑学硕士。本科以通识教育为基础。该模式既满足学科发展对高层次人才的需求，又满足了建筑行业对不同类型高层次人才的需要。基于该模式，建筑教育的目标可以由培养专业设计人才而上升到宽基础、善创新、高层次、国际化，引领整个建筑行业的高级人才。

2. 建筑学本、硕分类贯通的复合型教学框架。该教学框架分为：文理美通识、专业通识、专业提高与四大阶段，并以建筑设计为主轴，突出了专业特色。教学框架又分三个层次，分别对应了：建筑学行业专门人才（复合型）、建筑设计专业高层次人才（专业型）、建筑学科研究型高层次人才（研究型）。基于该框架，将通识教育、专业教育、专业提升和学术培养的分类教育理念落到了实处（图1）。

3. 以培养目标为导向、以知识类别为模块的课程体系。通过模块化课程的组织来实现不同人才培养路径，可以概括为：一条主干（设计课主干）、四个模块、多项选择。学术型人才——通识模块+专业模块+研究型模块；专业型人才——通识模块+专业模块+研究型模块；复合型人才——通识模块+专业模块+跨学科知识模块（图2）。

南京大学建筑教育的目标是培养建筑学高端人才，既要培养建筑设计的顶尖人才，也要培养建筑理论研究的顶尖人才，还要培养行业开发与管理的顶尖人才。但不管是"应用型"从事建筑设计工作，还是"学术型"从事建筑理论、科学技术的研究工作，还是"复合型"从事建筑行业的开发与管理工作，首先都需要对建筑设计有比较深入的了解，需要有很强的设计能力、创造能力与表达能力，也需要对建筑学整个学科有更深的理解。

二、比较：国内外现状与趋势简述

1. 国际现状

国外高水平的四年制建筑学本科的建筑设计课，以美国和欧洲的院校为代表。

美国麻省理工大学（MIT）采用四年制本科，第一年为新生建筑学研讨课，以seminar的形式上课，设计课从二年级开始加上四年级上学期，共五学期建筑设计课，学制与南京大学建筑学专业目前设计课程时间相同。这也是我院采用四年制本科体系中设计课程教学的一个重要参照。

瑞士苏黎世理工大学（ETH）学制为三年（高中为3+1）。由于其通识在高中的一年中完成，所以其实际上相当于四年制本科。瑞士苏黎世理工大学设计课共六个学期，课时逐年递增，一年级设计课每周6学时，第二年设计课每周10学时，第三年设计课每周16学时，即是在六学期内完成建筑设计课教学。

这两个学校代表新的教学体系，与我院目前实行的学制和教学体系十分接近，也代表了国际上建筑教育的某种新方向。

2. 国内现状

我国建筑学专业自1927年成立以来，"建筑设计课"作为主要专业课，教学模式主要来自第一代留学美国的杨廷宝等人带来的美国当时普遍实行的最早自巴黎美术学院的"图房"制，即师傅带徒弟的教学模式。差不多年来世界建筑学专业设计课最普遍采用的方式。这种方式重视绘图和图纸表达，而相对轻视技术和建造，教学方式是按照类型进行，不断重复，综合训练。国内最早成立的建筑院校普遍采用此方式，这种方式一直延续到现在。

国内高水平的建筑学专业普遍实行五年制本科，这种与注册职业建筑师制度和职业教育体系相关的学制，强调的是建筑学专业的职业教育，设计课程相应延长，加入了实习等环节，但建筑设计课的内容和教学方法并没有改变，还是沿用巴黎美术学院的体系。因此，建筑设计课，从中国近代建筑学诞生以来，其教学方式一直没有真正改变过。比如，国内最早获得教育部"建筑设计重点学科"的两所院校——清华大学"建筑设计专业"与东南大学"建筑设计专业"，均一直采用传统的教学方式，建筑设计课教学主要以类型主导、强调综合训练的方式。

3. 发展趋势

21世纪以来，在教育部"全国高等学校建筑学专业指导委员会"以及部分高校建筑学院的积极推动下，国内高水平的建筑院校如清华大学、同济大学开始实行4年制本科建筑学专业。清华大学部分实施四年制本科，设计课教学正在进行探索。同济大学2010年开始实施四年制模式，设计课程也在改革当中。在这一系列四年制改革中，也包含南京大学建筑与城市规划学院，其建筑学专业在2007年开始招收四年制本科生以来，一直致力于探索设计课的全面改革。对国际著名建筑学高校人才培养模式的调研，加上与国内建筑学专业教学计划的比较，发掘出一套与国际接轨的人才培养方案。南京大学建筑学将设计课教学设置在南京大学通识教育体系的框架下，利用综合性院校文理科优势，同时参照国际上一些著名高校的设计课程设置，在此方面做出了积极的尝试。

三、创新：建筑设计课模块化设置

以往多数中国高校建筑学五年制本科课程中，一年级开始美术课与设计基础课，二年级开始建筑设计课，以后建筑设计课程贯穿五年。而南京大学通识教育体系中前一、二年级以通识教育为主，专业课很少，三、四年级才开始有完整的建筑设计课，故此，学制缩短后建筑设计课如何保证质量就成为一个不可回避的重要问题。通识教育背景下四年制建筑学本科核心课程"建筑设计课"是重中之重，也以培养研究型人才，创建研究型大学，以及培养高端专业人才，创建高水平大学的重要环节之一。建筑设计课程设置成功与否，关系到通识教育和四年制建筑学专业本科的学制设置是否成功。

作为全国第一个建立在通识教育体系下的建筑学四年制本科专业，在此体系下的建筑设计课，必然要压缩设计课时间，调整教学方向，使教学计划更加紧凑，更加集约，教学目标更加精准。因此，南京大学建筑与城市规划学院把建筑学专业核心课程压缩为"设计基础I-II"（一年级）、"建筑设计基础I-II"（二年级）与"建筑设计I-VI"（三、四年级）三大课，其他课程设置均以此三大设计课为核心设置。具体举措为：

第一，以"设计基础I-II"代替"美术课"。以往建筑教育以素描、色彩（水彩或水粉）等传统美术课为主，训练学生观察、表达以及艺术修养，常常需要很长时间。现在以一门"设计基础"代替美术课，以现代设计教育思想代替传统美术，训练学生抽象与表达能力，同时培养观察表达能力，切入重点。

第二，以"建筑设计基础I-II"代替"建筑制图课"。以往建筑制图训练目标单一，仅关注训练制图，不关注建筑本身构成，不涉及学生对城市和建筑的理解。现在"建筑设计基础课"以认知基础，通过认知建筑、认知图示、认知环境、认知设计几个环节，强调认识建筑本身物质构成，以及其在环境中的意义，它不仅是一门制图课，还强调从理解建筑物与城市环境、城市环境开始学习建筑。

第三，以"建筑设计课I-VI"完成以往需要四年的全部设计课程。通过高度提炼的核心建筑问题为中心，组织六个设计题目。建筑设计一解决材料与建造问题，建筑设计二解决空间与表现问题，建筑设计三解决商业综合建筑的流线与功能问题，建筑设计四解决小区规划与住宅设计这个大量性问题，建筑设计五解决城市设计问题，建筑设计六解决高层建筑的规范和技术问题。

第四，专业课程设置围绕设计课展开，建筑原理、建筑技术课紧密结合设计课内容与进度，测绘、工地实习等利用假期进行。上课时间安排方面，设计课以及配套专业课集中在周二、周三，相对比较集中，学生有比较完整时间进行设计和讨论。

EXPLORATION ON THE SETUP OF UNDERGRADUATE DESIGN COURSES FOR THE DEPARTMENT OF ARCHITECTURE UNDER THE GENERAL EDUCATION BACKGROUND • ZHOU Ling DING Wowo

设计课程通过一系列高度集约化、体系化的设置，以建筑学基本问题出发，从材料—建造、空间—环境、结构—技术等为线索，递进式贯穿六个设计题目，涵盖了建筑设计中最重要的一些基本内容。通过设计课，学生不仅能学习建筑设计技能，还能学习相关技术与人文知识，更能训练创造性和扩展性思维。

四、实践意义

"建筑设计课"模块课程将成为南京大学建筑学科的核心课程，同时成为中国最早探索通识教育与建筑学结合的建筑设计课程。学生将通过课程训练掌握更加全面、实用的知识和技能。从知识传递、技能训练、创造力培养三个方面出发培养学生，改变了以往建筑设计课做不同建筑类型重复训练的只注重技能训练，而不注重知识传递与创造力培养的模式。

"建筑设计课"培养的人才将在各层次、类型上均达到一定的质量。学术型人才培养为具有宽厚基础与艰深学术研究能力的硕士与博士生输送人才，将胜任高等院校、科研及政府高管工作的人才；专业型人才培养为建筑学硕士生输送人才，将能胜任未来社会的建筑与城市重大工程的设计与设计工作；而复合型人才则为具有不同学科的知识技能交叉、复合的特色，适合为社会多种重要职位提供特殊人才。希望毕业生能以扎实的基本功、出色的研究能力和分析能力、出色的管理能力，获得行业内的高度认可。

在南京大学建筑学科人才培养分类贯通创新模式实验基地的条件下，南京大学建筑学科将实现在中国的建筑学教育领域中的宽基础、多层次、多类型的贯通式的人才培养，完成各类毕业生覆盖学术型、应用型和复合型的社会多种需求，从而促进中国建筑学教育的多元性和跨越性发展。

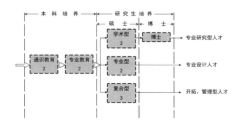

图1 通识教育体系下建筑学的培养目标

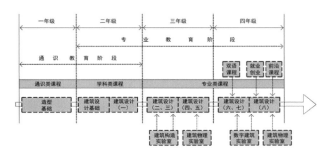

图2 四年制本科专业课程设置框架

1. Background: what kind of talents to develop

At present, although China's architectural education applies to the concept of internationally accepted professional degree education, the teaching mode is not completely in consistence. There are mainly three major problems: a. the degree in architecture is set repeatedly, and there is no clear division of undergraduate and graduate education, resulting in the overlong school system of the master's degree and the waste of educational resources. b. For a research-oriented university, the undergraduate education takes professional degree as the output, so it has to compress the general courses and enter the stage of professional training earlier, resulting in the limited space in the students' subsequent development. This status quo directly results in the lack of high-level research talent of undergraduates, making the architectural research and innovation lag behind the national high-end demand. c. Also due to the lack of the knowledge of general education, the students' academic horizons are narrower, far from satisfying the need of the whole architectural industry for diversified capacity, not to say meeting the new requirements of the country's future development.

To solve the above problems, the School of Architecture and Urban Planning of Nanjing University started the reform and innovation of architectural education mode in 2007, and carried out research and exploration on the concepts, methods and operational measures, mainly including the following aspects:

I. A new mode of the combination of general architectural education and professional education, namely: 4 Years (undergraduates) + 2/3 Years (postgraduates) mode. With reference to professional positioning among world-class universities, place the output of architectural professional education on the graduate level, and the students will receive a master of architecture degree. Undergraduates take general education as their foundation. This mode not only satisfies the needs of the discipline development for high-level talents, but also satisfies the needs of the architectural industry for different types of high-level talents. Based on this mode, the goal of architectural education can be elevated from the cultivation of professional design talents to high-level talents who have wide basis, are good at innovation, high-level and international, and will lead the whole architectural industry.

II. The composite architectural bachelor/master classification-through teaching framework. This teaching framework consists of three stages: Arts/Science/Aesthetics General, Specialty General and Specialty Advancement, with architectural design as the principal line, showing the professional characteristics. The teaching framework is divided into three levels, respectively corresponding to: special talents of architecture industry (versatile), high-level talents of architectural design (professional), and high-level research talents of architectural discipline (research). Based on this framework, we can put into practice the classified educational philosophy of general education, professional education, specialty advancement and academic training (Fig 1).

III. The curriculum system with training goals as orientation and knowledge categories as modules. Realize the training of different types of talents via modular course organization, which can be summarized as: one trunk (design courses trunk), four modules, and multiple options, i.e. academic talent — general module + professional module + research training; professional talent — general module + professional module + research type module; versatile talent — general module + professional module + interdisciplinary knowledge module (Fig 2).

The goal of architectural education of Nanjing University is to develop high-end architectural talents, including the top talents of architectural design, the top talents of architectural theory research, and the top talents of industrial development and management. However, for either the "applied talents" engaged in architectural design, or the "academic talents" engaged in architectural theory and science & technology research, or the "versatile talents" engaged in the development and management in the architectural industry, they need to have good understanding of architectural design, have strong design ability, creation ability and presentation skill, and have a deeper understanding of the whole discipline of architecture.

2. Comparison: summary on the status and trends at home and abroad

I. International status

Among the architectural design courses for the four-year architectural undergraduate in foreign countries, architectural schools in America and Europe are representative.

Massachusetts Institute of Technology (MIT) adopts the four-year undergraduate system. In the first year, the freshmen have architecture courses in the form of seminars; the design courses start from Grade 2, a total of 5 semesters of architectural design courses including Grade 2, Grade 3 and the first semester of Grade 4, of which the educational system is consistent with the schedule of current architectural design courses in Nanjing University. This is an important reference for us to teach the design courses in the four-year undergraduate system.

Eidgenossische Technische Hochschule Zürich (ETH) adopts the three-year system (3+1 for senior high school). Since the general education is completed in one year in the senior high school, it is actually equivalent to a four-year undergraduate programme. The design courses in ETH cover 6 semesters, with the class hour increased on a year by year basis: 6 hours a week in grade 1, 10 hours a week in grade 2, and 16 hours a week in grade 3. Similarly, it also completes the teaching of architectural design course in 6 semesters.

These two schools represent not only a new teaching system which is very similar to the current educational system and teaching system of our school, but also a new direction of the international architectural education.

II. Domestic status

Since the architectural major was established in China in 1927, the "architectural design course" has been primarily a professional course, and the teaching mode came from the atelier system from the École Des Beaux-Arts, i.e. a teaching mode of "master teaching apprentice", which was then very popular in the United States, and was brought in by Mr. Yang Tingbao and others, who were China's first-generation of overseas students in the United States. This is a way most commonly used over the years by architectural design courses around the world. This way emphasizes plotting and drawing representations, while relatively neglecting technology and construction; the teaching is carried out by building types, continuously repeating the comprehensive training. The earliest established domestic architectural institutions commonly adopt this way, which continues until now.

Since the 1990s, the domestic high-level architecture specialties have been widely using the five-year undergraduate system, which is related to the registered professional architects system and vocational education system, emphasizing the vocational education of architecture, with the design courses accordingly extended and internship added. However, the content and teaching method of architectural design courses remains unchanged, sticking to the teaching system from the École Des Beaux-Arts. Therefore, since the birth of the modern Chinese architecture specialty, the teaching method of architectural design course has not really changed. For example, for the two institutions who were first to achieve the "key discipline of architectural design" of the Ministry of Education — Tsinghua University and Southeast University, their "architectural design specialty" has been using the traditional teaching system, and the teaching of architectural design courses is led by building types, emphasizing comprehensive training.

III. Trends

Since the 21st century, with the active promotion of the Ministry of Education's "National Steering Committee for College Architecture Specialty", as well as some of the architectural schools, the high-level domestic architectural institutions, such as Tsinghua University and Tongji University, have started to implement a four-year undergraduate architectural programme. Tsinghua University partially implements the four-year undergraduate system, and explorations on design course teaching are under way. Tongji University started the implementation of the four-year system in 2010, and the reform of the design course is also under way. This series of four-year system reform also includes the School of Architecture and Urban Planning, Nanjing University, which started to recruit four-year system undergraduates in the architecture specialty in 2007 and has been committed to exploring the design courses. Based upon the survey on the talent training model of the internationally renowned architecture schools, coupled with the comparison with the teaching plans of domestic architecture specialty, the School of Architecture and Urban Planning, Nanjing University, has developed a set of internationally compatible talent training programs. The architectural discipline of Nanjing University sets up the design course teaching under the framework of the Nanjing University general education system, to make full use of the arts and science advantages of a comprehensive university, and consulting the design course settings of some internationally renowned universities, it has made positive attempts in this aspect.

3. Innovation: Modular setting of architectural design courses

In the previous five-year architectural undergraduate courses in most Chinese colleges and universities, fine arts and design basic courses started in grade 1, architectural design courses started in grade 2 and will continue to be taught throughout the five years. In Nanjing University's general education system, teaching in grade 1 and grade 2 focuses on general education, and complete architectural design courses will be available in grade 3 and grade 4. Therefore, how to ensure the quality of architectural design courses after the compression of the training system has become an unevadable key issue. Under the general education background, the four-year architectural undergraduate kernel course "architectural design" is a top priority, which is also one of the important aspects for training research talents, creating research universities, developing high-end professionals and creating high-level universities. Whether the architectural design curriculum is successful relates to the success of the general education and the four-year architectural undergraduate programme.

As China's first four-year undergraduate architecture specialty established under the general education system, the class hours of architectural design courses have to be compressed and the teaching direction needs to be adjusted to make the teaching plan more compact, more intensive, and the teaching objectives more precise. Therefore, the School of Architecture and Urban Planning of Nanjing University has compressed the core architecture courses into "design basis I-II" (grade 1), "architectural design basis I-II" (grade 2) and "architectural design I-VI "(grade 3 and grade 4), and all other curricula take these three design courses as core curricula. The specific initiatives include:

I. Replace the "fine arts" with "design basis I-II". The previous architectural education mainly focused on traditional fine arts courses such as sketch, chromatics (watercolor or gouache) to train students' observation, expression and artistic accomplishments, often taking a long time. And now, replacing the courses on fine arts with a "design basis" course and replacing the education of traditional fine arts with modern design education the new programme trains the students' abstract expression skills and meanwhile develops their observation skills, thus proceeding to the key point.

II. Replace the "architectural drawing" with "architectural design basis I-II". The previous architectural drawing has a single target, concerned only with training in drawing rather than architecture per se, and has no relation to the students' understanding of the city and architecture. Now, the "architectural design basis" course is based on understanding. By understanding architecture, understanding graphic representation, understanding built environment and understanding design, the course stresses the awareness of the material composition of architecture and its significance in the environment. It is not only a drawing course, but also stresses learning architecture from understanding the building and built environment as well as urban environment.

III. Complete all design courses previously requiring 4 years by taking the course "architectural design I-VI". Organize six design themes by centering on the highly refined core architectural issues. Architectural design I deals with the materials and construction issues, architectural design II deals with the space and representation issues, architectural design III deals with issues of circulation and function in the case of commercial building complex, architectural design IV deals with the issues of residential quarter planning and housing design, architectural design V deals with the urban design issues, and architectural design VI deals with the specifications and technical issues of high-rise buildings.

IV. The specialized curriculum expands around the design course, the architectural principles and architectural technology courses closely integrate with the content and progress of the design course, and carries out surveying, mapping and site practice during the vacations. As for the class hour arrangements, the design course and its supporting professional courses are allocated to Tuesday and Wednesday, which are relatively concentrated, thus giving students more time to conduct design and discussion.

Through a series of highly intensive and systematic curricula, the design courses start from the basic architectural issues, taking materials / construction, space / environment, structure / technology as clues to progressively link to the six design topics, covering the most important fundamental contents in architectural design. The design courses train the students to learn the architectural design skills and the related technical and humanistic knowledge, and to train their creative and extensive thinking.

4. Practical significance

The "architectural design" module will become the core course of architecture in Nanjing University, as well as China's first architectural design course that explores the combination of general education and architecture. The students will master more comprehensive and practical knowledge and skills through the course training. It will develop students from three aspects including knowledge transfer, skill training, and creative ability development, and change the previous model in which the architectural design course focuses only on the skill training by repeating the exercise on different architectural types rather than focusing on knowledge transfer and creative ability development.

The talents trained by the "architectural design" course will achieve a certain quality in various levels and types. The academic talents will be trained to be master's and doctoral talents with a wide basis and stronger academic research abilities, who will be qualified for senior management in colleges, scientific research institutions, and the government; the professional talents will be trained to be master of architecture, who will be qualified for the design and research of architecture and key urban engineering; while the versatile talents feature having crossed and composite knowledge and skills of different disciplines, applicable for providing special talents for various important positions in the society. The graduates are expected to obtain high acceptance by the industry for their solid basic skills, excellent research and analyzing abilities as well as sound management power.

Under the condition of Nanjing University's experimental base for the innovative model of classification-through architectural talent development, Nanjing University's architecture discipline will implement the wide base, multi-level and multi-type through-type talent development in the field of China's architectural education, and satisfy diversified social demands for various types of graduates covering research talents, applied talents and versatile talents, thus facilitating the diversified and leap-frog development of China's architectural education.

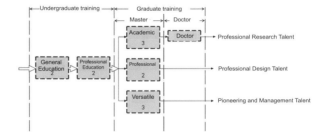

Fig.1 Architecture training objectives under the system of general education

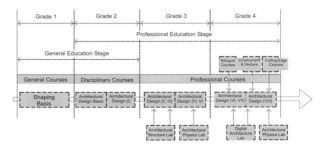

Fig.2 Framework of the specialized courses for four-year undergraduates

南京大学外国建筑史教学经验谈 • 胡恒

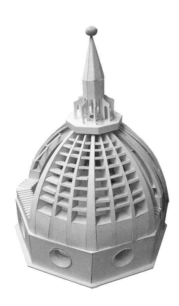

建筑史课程与建筑设计课程的关系,一直都是本科教学的一个问题。一般来说,建筑史教学有大量内容相关于对作品的讲解,尤其是现代建筑部分,这无疑与建筑设计课程颇多关联。在涉及柯布西耶、密斯,以及当代的建筑作品的时候,这些知识对于设计的重要性更显突出。在较新版本的《近现代外国建筑史》教材中,编辑组加入大量的库哈斯、安滕忠雄以及瑞士与西班牙建筑师的新旧作品,其意义也在于此。可以说,在某种程度上,建筑史教学是设计课程的辅助。

南京大学的外国建筑史教学中,建筑史课程与设计课程之间有一个主动的分离。它不需要为建筑设计提供历史知识。它是一个关于独立知识系统的讲解机制。这一种独立性产生于两个原因:

第一,建筑设计课程已成独立架构。互联网的繁荣和书籍杂志刊物的普及,使得学生对建筑设计的历史知识的获得相当便捷。而对当下建筑设计的全球动向和资讯的掌握,学生的渠道更为宽广。并且南京大学建筑设计教学的体制经过多年实验,已渐趋完善。各个因素加在一起,使得建筑设计课程的体系基本自主。相对应的,建筑设计课程已不再需要建筑史课程作知识支撑,两者之间传统的主辅关系解体。这无疑取消掉建筑史课程旧有的责任与合法性,动摇了其存在的基础,给建筑史教学带来危机。

第二,建筑史课程本身亦有独立架构的可能。建筑史教学的危机,也是迫使我们重新思考其学科定位、核心价值、知识构成、教学模式等一系列的问题的一个机会。外国建筑史和中国建筑史不同。中国建筑史是一种主观的知识,因为对它的讲解和学习,是伴随着自身的体验来进行的,并且这一知识的传统是我们祖先所创造和流传,更使得它内在于我们的主体。所以,无论是学习或是进一步的研究,它对我们而言并无本质上的距离,只在于有无介入的必要和介入深度的问题。相反,外国建筑史是一种客观的知识。它离我们遥远又陌生。我们既无法对建筑本体获得足够多的身体感知,也对产生它们的社会背景与文化根源知之甚少。这种距离感,确实已经威胁到外国建筑史教学的根基,但这也是使该知识形式在课堂上彻底客观化的契机。

客观化的方式为:确定建筑史的文化特征,将建筑史归还到文化史范畴;确立人(建筑大师)对建筑史的意义——他们的作品序列构成独立完整的历史段落。前者划定了教学的大范围(第一学期的课程);后者划定教学的小区间(第二学期的课程)。两个课程建构起建筑史教学的核心价值——文化对建筑的滋养,以及人对于建筑的意义。

我们第一学期的"通史教学"是将建筑放置在文化整体性里来考虑的。虽然讲解的段落按历史时间而定——希腊、罗马、中世纪、文艺复兴、巴洛克、新古典主义……但是,我们并非依照建筑风格的区分来展开,而是按照不同时期的综合文化特点来为建筑寻找相应的位置。比如希腊时期,我们的讲解重点不在神庙的形制与柱式,而是在地理形态、文明冲突、宗教生活、文学与雕塑艺术等多种文化要素与自然要素的多向作用上。正是在这个综合作用的氛围中,神庙的雕塑观念的指向、形制的流变才显现出一种必然性与合理性。再比如古罗马时期,希腊时期的讲解重点全然有变。在此,决定建筑空间类型变化(转向内空间的塑造)的因素是罗马诸任皇帝个体趣味的差异、罗马式"天下观"的成型、罗马与埃及两种文化的交互作用、市民娱乐生活的兴起、城市尺度。文化的整体性是"通史教学"的基本平台,它将客观知识——文明的冲突、地理环境、大事件、日常生活与艺术、材料与技术——编织成一张宽大厚重的网,使建筑能够恰当地落在一个具体且坚实的基础上。

"通史教学"将建筑的专业性分解为若干文化要素的共同决定论。对这些要素的掌握无需复杂的专业准备。它们具有某种普遍性,并且容易理解,可以轻易唤起学生的共鸣。通过教学和学习,学生会留下一个意识:建筑并非抽象图示的进化,或者纯

ON NANJING UNIVERSITY'S EXPERIENCE IN FOREIGN ARCHITECTURAL HISTORY TEACHING • HU Heng

技术的表现，而是文明的产物，是多种社会因素共同作用的结果。它的专业性体现在与特定时代、特定环境的不可分割的紧密联系中。这亦使之成为可理解之物，而非艺术形式的某种神秘衍生过程。

这一学期也是南京大学学生初次正式接触建筑的学期（第一年是全校的通识课程）。外国建筑史的课程希望给建筑铺展开为一个广阔的世界，而不仅仅是一项专业技能的演变过程。

第二学期的课程（个案研究教学），是"通史教学"的后续阶段。在通史课中，文化的整体作用是主基调，人（建筑师）的形象不是很重要。这里谈及的建筑大多是匿名的。从希腊神庙到罗马浴场，再到哥特大教堂，皆是如此。在个案教学阶段中，人的意义是主导性的。也即，这门课的主题是建筑师——16周（一周2课时）的课程，分为16个单元，1~2个单元讲解一个建筑师，课程一半为文艺复兴建筑师，一半为现代主义建筑师。

在我看来，建筑师的时代以意大利文艺复兴与第二次世界大战前后的现代主义最有代表性。在这两个时代中，建筑大师以星群的方式出现。这是天才爆炸的时代。建筑师以个人的名义推动建筑史的发展、改变建筑文化的走向，同时也创造了建筑学学科的价值内核与基础。所以，尽管这两个时期的长度有限（前者约200年，后者约50年），但理所当然是我们教学的重点。

我们选取的文艺复兴建筑师为伯鲁乃涅斯基、阿尔伯蒂、伯拉孟特、米开朗琪罗、罗马诺、桑索维诺、帕拉迪奥，现代主义建筑师为高迪、赖特、密斯、柯布西耶、海杜克、路易康。这些建筑师的作品，我们并非挑选几个代表作来做简略的介绍，而是尽力在课程中铺陈其全部的作品。这种全景式的作品描述很繁琐。比如伯鲁乃涅斯基的作品，我们既要详细讲解佛罗伦萨主教堂的穹顶、育婴堂、拜奇礼拜堂（这是教材里有的），还需谈及他在佛罗伦萨的两个较大的教堂（圣洛伦佐教堂和圣灵教堂）、大教堂的采光亭、圣玛利亚·德·安杰利教堂方案、归尔甫宫、剧场装置设计。尤其是最后几个并不太重要的作品，它们没有出现在教材中，一般的历史书也甚少强调。但是它们都是课程讲解的不可缺少的对象。

全景式的描述的作用有两个方面。一方面，它们可以让学生领会到，作为一个建筑大师，在其一生中能有多少作为——他们的努力都在建筑上深深刻下自己的名字。尽管在希腊神庙、哥特教堂都有建筑师的名字留下来，但是建筑师要成为有丰满形象与独立社会意义的个体，则要在文艺复兴时期才得以慢慢成形。并且他们的社会地位的提高，和相关的理论著作的不断面世也使得建筑师作为一门职业与建筑作为一门学科在此出现。学生是以独立个体的形式进入学习之中，学习对象以"人"为中心，这可以建立起有质感的参考系——这是一种伴随终身的潜移默化。

另一方面，全景讲解，可以展开多种重要的主题。这些主题在建筑大师身上同时出现，但是它们朝向不同的方向，在其他建筑师的作品中得到不同程度的延续和深化。关于这些主题的阐述，也是文艺复兴建筑史教学的重点。它可以帮助学生将不同的建筑之间的联系建立起来。文艺复兴建筑史就是由很多线索、网络、系统的不同层面的叠加。比如伯鲁乃涅斯基的几个不太显眼的作品，它们虽然在艺术成就方面不如那些更为著名的作品，但是都涉及文艺复兴时期的重要建筑主题。

安杰利教堂方案虽然只有方案留下来，但是它是第一次对教堂的集中式形制所做的尝试。在稍后的阿尔伯蒂、伯拉孟特、米开朗琪罗的教堂建筑中，该实验都有着不断的推进。可以说，罗马宗教建筑的主要形制如何转换到基督教教堂建筑中，这一文艺复兴建筑师都面临的重大建筑命题，在伯鲁乃涅斯基的这个未完成的方案里首开先河。其开端的意义毋庸置疑。

归尔甫宫虽然在设计上并无多少出彩之处，但是它也是一种重要建筑类型（府邸建筑）的开端。在米开罗佐的美第奇府邸、阿尔伯蒂的鲁切莱府邸、罗塞蒂的斯特罗奇府邸中，都能看到归尔甫宫开辟的道路。另外，这些佛罗伦萨的府邸建筑与当地的银行家族关系密切。归尔甫宫的意义还在于我们对建筑师与赞助人之间的关系的理解。这些关系是建筑师作为个体的人与社会联系的体现。通过对这些人与人之间的关系的认知，学生对这些府邸建筑的形式的特点、差异的理解自然会有章可循。

剧场装置设计，是伯鲁乃涅斯基的一个计划。它没有图纸和模型留下来。其信息只在瓦萨里的《名人传》中有一段较为细致的描述。这个剧场设计虽然信息极为缺乏，但是它亦有特殊的意义。和安杰利教堂、归尔甫宫一样，这是文艺复兴的一种建筑形制——舞台设计——的一个重要的开端。该形制是建筑的公共空间开始成为建筑师创作对象的象征。在经过塞里欧的《建筑七书》等著作的过渡之后，它在文艺复兴的最后一位大师帕拉迪奥的奥林匹亚剧场的设计中达到顶峰。

以上几个主题（集中式教堂、府邸建筑、建筑师与赞助人、剧场设计）都集中在伯鲁乃涅斯基一人身上。全景式讲解可以通过一个点来展开多种重要的主题方向。这既可以加重学生对"人"的认识（以及各个文艺复兴建筑师之间的传承关系），也可以对涉及的诸多建筑主题都有所了解。

由于这些个案建筑师的作品分量都相当大——在文艺复兴的最后个案帕拉迪奥那里，由于其作品实在太多（分维琴查时期、小住宅系列、威尼斯的教堂三部分），尽数展开需要花费3次课程的时间。所以，在这学期的有限的32个课时中如何分配不同建筑师的份额，也是需要斟酌的问题。

The relationship between architectural history courses and architectural design courses has always been a problem in undergraduate teaching. In general, the teaching of architectural history is closely related to the interpretation of design projects, especially in the section of modern architecture. This is undoubtedly associated with many architectural design courses. The importance of this association becomes even more evident when involving Le Corbusier, Mies van der Rohe, as well as contemporary architectural works. This is why the editorial committee added the works of Rem Koolhaas, Tadao Ando, and Swiss and Spanish architects in the newer version of the textbook "Modern and Contemporary Foreign Architectural History". It can be said that, to a certain extent, architectural history teaching is an aid to the design courses.

However, in the foreign architectural history teaching of Nanjing University, there is a deliberate separation between the architectural history courses and design courses. It does not intend to supply architectural design with historical knowledge. It provides a mechanism of interpretation for an independent knowledge system. This kind of independence is due to two reasons.

Firstly, the architectural design courses have formulated an independent framework. The booming of Internet and the popularity of the publication of books and magazines facilitate the students' acquisition of historical knowledge about architectural design. The grasp of the global trends and information on current architectural design brings students broader channels. In addition, the architectural design teaching system in Nanjing University has become more complete after a decade of experiments. A combination of these factors makes the architectural design teaching system in Nanjing University more autonomous. Correspondingly, architectural design courses no longer requires the knowledge support of architectural history courses. The traditional primary-secondary relationship between them has been dismantled. This has undoubtedly nullified the traditional responsibility and legitimacy of the architectural history courses, undermined the bases for their existence, and brought crisis to the teaching of architectural history.

Secondly, the architectural history courses also have the possibility to have their own framework. The crisis of architectural history teaching also provides an opportunity that forces us to rethink a series of issues such as the positioning of the discipline, its core values, knowledge constitution, and mode of teaching. Different from foreign architectural history, Chinese architectural history is a kind of subjective knowledge, because the interpretation and learning of it is accompanied by our own experience. And the tradition of this knowledge is created and disseminated by our ancestors, making it part of our spirit. So either for learning or for further research, it has no essential distance for us, and what matters is only the necessity of intervention and the depth of intervention. In contrast, foreign architectural history is a kind of objective knowledge. It is far away from us and is unfamiliar to us. We can neither acquire enough bodily experience of the architecture itself, nor know much about its social background and cultural roots. This sense of distance does threaten the foundation of foreign architectural history teaching, but it also provides an opportunity for the thorough objectification of such a form of knowledge in the classroom.

The way of objectification is: determining the cultural characteristics of architectural history and returning architectural history to the scope of cultural history; establishing the significance of individuals (master architects) to the architectural history — the sequence of their works constitutes independent and complete historical segments. The former determines the broad scope of teaching (courses in the first semester); the latter determines the micro-segments of teaching (courses in the second semester). Both courses construct the core values of architectural history teaching — on culture's nourishment of architecture, as well as the significance of people to architecture.

The general history teaching used in the first semester was to consider architecture within cultural integrity. Although the order of introduction was based upon historical timeline — Greek, Roman, Medieval, Renaissance, Baroque and Neoclassicism, we didn't unfold the discourse by the architectural styles, but searched for the appropriate position for architecture in accordance with the cultural characteristics of different periods. For example in the Greek period, our explanation didn't focus on the layout and orders of the temples, but focused on the multidirectional forces of cultural and natural factors, such as topography, the clash of civilizations, religious life, literature and sculptural art, so on and so forth. It is in this atmosphere of integrative forces, the developments in the ideas of sculpture and the changes in the layout of temples showed a necessity and rationality. Another example is the Roman period; the explanation focus of the Greek period is completely changed. Here, the factors to determine the type of architectural space changes (turning toward the creation of the inner space) are the individual taste differences between all Romal emperors at different terms of office, the formulation of a Roman idea of the "world", the interaction of Roman and Egyptian cultures, the rise of public entertainment and the urban scale. The cultural integrity is the basic platform for "general history teaching". It weaves objective knowledge — the clash of civilizations, geographical environments, big events, daily life and arts, materials and technology — into a wide and thick network, enabling architecture to appropriately stand on a concrete and solid foundation.

The "general history teaching" decomposed architectural disciplinarity into the common determinism of several cultural factors. To master these factors does not require complex professional preparation. They have some sort of universality and are easy to understand, can easily arouse the students' resonance. Through teaching and studying, the students will obtain a consciousness: architecture is not the evolution of abstract icons or pure technical expressions, but a product of civilization, an outcome of the interaction of a variety of social factors. Its professionalism is reflected in the inseparable contact with the specific era and environment. This also makes it an understandable thing, rather than a certain mysterious evolution process of the art form.

This is a semester when the Nanjing University students make their first formal contact with architecture (in the first year, all students of the university are taught in general education courses). The foreign architectural history courses are expected to spread out a vast world for architecture, rather than the evolution process of a professional skill.

The course for the second semester (Case Study Teaching) is the subsequent phase of the "General History Teaching". In the General History courses, the overall role of culture is the main tone, and the performance of the person (architects) is not very important. The architecture mentioned is mostly anonymous. This is true for

Greek temples, Roman Baths, and Gothic cathedrals. At the case teaching stage, the significance of the person becomes dominant. That is, the theme of this course is the architect — 16 weeks' (2 lessons a week) courses are divided into 16 units, explaining one architect every 1~2 units. In the courses, Renaissance architects and modernist architects occupy 50% respectively.

In my opinion, the "age of architects" is best represented by the Italian Renaissance era and before and after World War II. In these two eras, the master architects appeared as constellations. This is an era of the explosion of genius. The architects personally promoted the development of architectural history and changed the direction of architectural culture, while creating the value kernel and basis of the architectural discipline. Therefore, despite the limited length of these two periods (about 200 years for the former, about 50 years for the latter), they are of course the foci of our teaching.

The Renaissance architects in our selection include Filippo Brunelleschi, Leon Battista Alberti, Donato Bramante, Michelangelo Buonarroti, Giulio Romano, Jacopo Sansovino and Andrea Palladio, and the modernist architects include Antoni Gaudi, Frank Lloyd Wright, Mies van der Rohe, Le Corbusier, John Hejduk and Louis Kahn. Among the works of these architects, we do not pick a few representative works to give a brief introduction, but try to lay out all their works in the curriculum. This kind of panoramic description of works is very cumbersome. For example, for the works of Filippo Brunelleschi, in addition to giving a detailed explanation to the dome of the Florence Cathedral, Foundling, Babich Chapel (these are included in the textbooks), we still need to talk about his two larger churches in Florence (Saint Lorenzo and Saint Spirito), the lighting lantern of the Cathedral, the project for the Santa Maria degli Angeli Church, and Palazzo di Parte Guelfa, and the design for theatric device, especially the last few less important works. They didn't appear in the textbooks, and were seldom stressed in general history books. However, they are indispensable objects in the course.

The panoramic description has two roles. On one hand, they can enable the students to understand how many efforts a master architects can make in their life time — their efforts carved their names deeply into architecture. Although many Greek temples and Gothic Churches have kept the name of their architects, professional architect, as a subject with a full image and independent social significance, was only gradually shaped in the Renaissance period. In addition, the improvement of their social status and the continuous publication of relevant theoretical treatises facilitated the emergence of the architect as a career and architecture as a discipline. The students enter learning as individual subjects, and their learning object is centred on the "person", which establishes a reference system with texture — providing an imperceptibly influence over their lifetime.

On the other hand, in the panoramic description, you can unfold a variety of important topics. These topics also appear in several master architects. However, they are continued and deepened to different degrees in the works of other architects. The elaboration on these topics is also the focus of the teaching of Renaissance architectural history. It can help students set up a link between different architectural works. The Renaissance architectural history is superimposed by many clues, networks, and systems at different levels. For example several less prominent works of Filippo Brunelleschi, although their artistic achievements are less evident than those more famous works, all of them are related to the important architectural themes in the Renaissance period.

Although the project of Santa Maria degli Angeli Church only left the proposal to us, it is the first known attempt made towards central-plan church. Among later church architecture of Alberti, Bramante and Michelangelo, this experiment was constantly advancing. It can be said that how had the primary form of Roman religious architecture been converted into Christian church architecture — a major architectural issue facing the Renaissance architects, was first experimented in Filippo Brunelleschi's proposal. Its significance as a beginning is unquestionable.

Although the Palazzo di Parte Guelfa doesn't have many innovative features in design, it is also the beginning of an important architectural type (mansion building). In Michelozzo's Medici Mansion, Alberti's Palazzo dei Rucellai Mansion and Rossetti's Palazzo Strozzi Mansion, we can find the way opened up by the Palazzo di Parte Guelfa. In addition, these Florentine mansion architecture were closely tied with local banking families. The significance of the Palazzo di Parte Guelfa also lies in our understanding of the relationship between the architects and their sponsors. These relationships manifest the social ties of the architects as individuals. By understanding these relationships between people, the students will naturally have rules to follow in the understanding of the characteristics and differences of these forms of architecture.

Theater device design, a plan of Filippo Brunelleschi, did not leave any drawings and models to us. It only appeared in a detailed description given by Giorgio Vasari in his book *Le Vite de' più eccellenti pittori, scultori, ed architettori*. This theater design, although with only scarce information, is of special significance. Similar to the Santa Maria degli Angeli Church and the Palazzo di Parte Guelfa, this is an important beginning of a type of Renaissance architecture — stage design. This type symbolizes that the public space of architecture starts to become the object of architects' creativity. After a transition of Sebastiano Serlio's *Seven Books of Architecture* and other works, it peaked in the last Renaissance master Palladio's design of Olympia Theater.

The above topics (central-plan church, mansion building, architect and sponsor, theater design) are concentrating on Filippo Brunelleschi. The panoramic presentation may unfold a variety of important thematic directions from one point. This can not only intensify the students' understanding of the "person" (as well as the inheritance relationship between various Renaissance architects), but also inform them of many involved architectural themes.

Due to the quantity of the works of the case architects, e.g. with the last case in the Renaissance period, Andrea Palladio, who has an exceptional large amount of works (sorted into the Vicenza period, a series of country residences, and Venetian churches), it will take three classes to exhaust all of them. So, how to allocate the lectures on different architects within the limited 32 class hours is also a problem to be solved.

设计本身节能的健康建筑 • 秦孟昊

什么样的建筑才能令人真正难忘？是建筑杂志上光鲜靓丽的效果图还是令人兴奋的沙盘模型？我想真正令人难忘的应该是人们在这栋建筑中居住、工作过程中的舒适感受，以及人们在建筑空间中穿行时愉悦的体验。

自然环境中的热、光、声等因素对一座建筑的体验美学、建筑运行、室内环境有着至关重要的影响。综合考虑这些因素对建筑能耗、舒适性以及健康方面的影响应该成为建筑师在建筑设计最初阶段就开展的主动行动。建筑师不仅要考虑建筑物内部各种使用功能和使用空间的合理安排，建筑物与周围环境、与各种外部条件的协调配合，内部和外表的艺术效果，各个细部的构造方式还需要考虑建筑设计对建筑能耗和室内环境质量的影响。其最终目的是使建筑物做到节能、健康、适用、经济、坚固、美观。

为了达到以上的要求，建筑师需要学习掌握绿色节能建筑设计的基本原理、基本工具、重要数据和常用计算方法。只有这样才能使所设计的建筑既美观又舒适、节能、经济，并使人们在建筑中流连忘返。

南京大学建筑与城市规划学院从自身的具体情况出发不断探索在建筑学教育中推广深入建筑技术科学专门化教育的尝试，其目的就是培养建筑学学生主动关注建筑节能与环境健康的观念，学会绿色节能建筑的基本设计方法，并学会运用各种工具完善建筑设计方案。具体的课程设置安排如下：

建筑技术课程建设一览（本科）
建筑环境类课程
人居环境与可持续发展
环境科学导论
建筑物理（声、光、热）
建筑设备（水、暖、电）
建筑节能与绿色建筑设计（高年级、综合性生态与节能）
建筑设计专题（节能与生态专题设计，四年级本科毕业设计）
数字建筑与建造类课程
建筑构造
CAAD理论与实践
BIM技术运用
国家大学生创新计划

建筑技术课程建设一览（研究生）
主要课程
现代建筑技术
建筑体系整合
传热学与计算流体力学基础
CFD与建筑环境设计
绿色建筑虚拟设计平台VDS
高等传热学（建筑环境专业同学选修）
高等工程热力学（建筑环境专业同学选修）
硕士论文

南京大学建筑与城市规划学院可持续建筑研究中心

南京大学可持续建筑研究中心成立于2010年，主要从事可持续建筑、建筑节能，特别是长江中下游地区湿热（冷）环境下建筑设计与节能分析方面的理论与应用研究。目前，中心下设三个研究方向（梯队）：整体建筑热湿耦合传递研究、建筑采光节能技术和可持续建筑设计与教育。

中心现有专/兼职人员12人，固定成员包括：教授2名，副教授3名，讲师1名，所有成员均具有博士学位，并都有国外学习、工作经历，成员具备先进的知识和视野。近年来，中心承担十多项科研任务，主要包括国家"863"课题1项，国家自然科学基金5项以及多项省部级项目和国际项目。

中心建有建筑环境实验室，实验室现有面积500余平方米，已建成包括大型高精度人工气候室在内的一批建筑节能、环境和设备测试试验台，购置了大量的建筑节能、建筑环境测试仪器，可以为中心的科学研究和社会服务提供良好的支撑。

中心重视对外学术交流和研究平台建设。近年来，先后与美国Syracuse大学、美国佛罗里达太阳能中心、美国国家标准局、英国女王大学、英国剑桥大学、法国国家建筑物理实验室、瑞士苏黎世理工大学、瑞典隆德大学等10余所国际知名的大学和科研院所建立了合作关系。

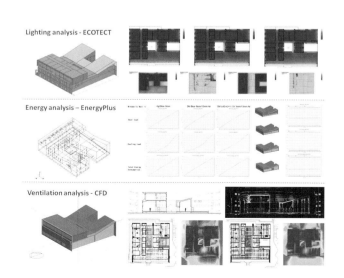

DESIGN FOR GREEN AND HEALTHY BUILDING • QIN Menghao

What makes buildings truly memorable? The glossy design sketch on the architectural magazine or the exciting scaled model? In my opinion, it comes from the pleasant experiences of people who live, work and pass through the interior spaces. The heat, light, sound and other factors in the natural environment have a strong impact on a building's experiential aesthetic, operation, and indoor environment. Considering these factors' impact on the building's energy consumption, comfort and health should be a purposeful action of architects in the initial stage of design. The architect must consider not only the reasonable arrangement of various use functions and use spaces within a building, the coordination with the surrounding environment and various external conditions, and the artistic effect of the interior and exterior and the structuring method of all details, but also the architectural design's impact on building energy consumption and indoor environment. Its ultimate goal is to make buildings energy-efficient, healthy, practical, cost-effective and beautiful.

To achieve the above requirements, architects need to learn and master the basic principles, basic tools, important data, and calculation methods of green and energy-efficient architectural design. Only in this way, can make architecture beautiful, comfortable, energy-saving and cost-effective; and allow people to enjoy themselves so much as not to leave the building.

Based on its own specific conditions, the School of Architecture and Urban Planning of Nanjing University keeps exploring the attempts of promoting specialized education in architectural technology during the education of architecture, aiming at training the architectural students' perception of actively paying attention to the building energy-saving and environmental health, to learn the basic design methods for green and healthy building, and to know how to use various tools to improve the architectural design. The specific curriculum is as follows:

List of architectural technology curriculum development (undergraduate)
Courses of built environment
Habitat environment and sustainable development
Introduction to environmental sciences
Building physics (sound, light, heat)
Building equipment (water, heating, electricity)
Building energy efficiency and green building design (for higher grade, comprehensive ecological and energy-saving)
Specialized architectural design themes (energy-saving and ecological themes design, for four-year undergraduates' graduation design)
Courses of digital architecture and construction
Building construction
CAAD theory and practice
BIM technology application
National university students innovation program

List of architectural technology curriculum development (postgraduate)
Main courses
Advanced building technology
Integration of architectural system
Basis of Heat transfer and computational fluid dynamics
CFD and architectural environment design
Virtual design studio for green architecture - VDS
Advanced heat transfer (elective for students of built environment)
Advanced engineering thermal dynamics (elective for students of architectural environment)
Master thesis

Center for Sustainable Building Research, School of Architecture and Urban Planning, Nanjing University
Established in 2010, the Center for Sustainable Building Research in Nanjing University is mainly engaged in the theoretical and applied study of sustainable architecture and building energy efficiency, especially the green building design and energy performance analysis under hot(cold) and humid climate in the middle and lower regions of the Yangtze River. At present, the center has three research directions: whole building coupled heat and moisture transfer, high performance energy-efficient building techniques, and design and education of sustainable buildings.

The center has a total of 12 full-time faculty and staff, ,which include 3 professors, 3 associate professors, and 1 lecturer, all of whom have a PhD degree and advanced knowledge and vision, and have the experiences of studying or working abroad. In recent years, the center has undertaken dozens of scientific and research tasks, mainly including 1 national "863" project, 5 national natural science foundation funded projects, and a number of provincial and ministerial projects as well as international projects.

The center has a built environment laboratory with an area of over 500 square meters, has completed a number of advanced test facilities for the research of building energy efficiency, indoor environment, and moisture transfer, which include a large high-precision artificial climate chamber. The center has purchased a large number of building energy efficiency and built environment test equipment, being capable of providing good support for the center's scientific research and social service.

The center gives great importance to foreign academic exchanges and cooperation. In recent years, it has successively established cooperative relationship with over 10 world-famous universities and research institutes, including Syracuse University, Florida Solar Energy Center, the U S National Institute of Standards and Technology (NIST), Queen's University, the University of Cambridge in the UK, The University of La Rochelle (France), ETH (Switzerland), Lund University (Sweden), etc.

建筑技术课程的创新探索 • 吴蔚

1969年，Reyner Banham在他的论著 *The Architecture of the Well Tempered Environment* 的第一章 "Unwarranted apology" 中就强调到，他的这本书讲的是建筑历史和理论，而不是介绍建筑技术[1]。然而令人遗憾的是，43年后的今天，他的这本书还是放在全世界各大建筑图书馆中建筑技术一栏中，建筑技术与建筑设计无论是在实践上还是教学方面都是越走越远。无可否认，建筑技术对现代建筑设计和理念的影响是前所未有的，建筑技术也是建筑学教学中不可缺少的必要环节。但尽管建筑技术方面的课程如"建筑物理"、"建筑设备"等都是建筑学本科的必修课，却都被"公认为"最不为学生所重视的专业基础课。如何将建筑技术的理论知识与建筑设计更好地相融合，不仅是国内外建筑学专业在教授建筑技术课程上面临的主要问题，也是建筑设计实践中很多建筑师在探讨的热点问题之一。

尤其是近十几年来，建筑的可持续发展在建筑设计领域占有越来越重要的地位，以及我国对建筑节能的日益重视，建筑师对与技术相关的知识的需求量日益增大，建筑技术课程面临着新的机遇和挑战。建筑技术课程需要改革和创新，使建筑学学生不仅更好地掌握传统的建筑技术基础知识，还需要革新教学内容，使之与时俱进，让学生将所学的建筑技术融会贯通到建筑设计中。

本文主要介绍南京大学建筑与城市规划学院建筑学三年级本科两门技术课程的教学改革理念和探索，总结和分析笔者在革新实践中所遇到的困难和问题。本文旨在帮助我国的建筑学院、系在改革创新建筑技术课程时，能够让学生更好、更快地掌握建筑技术方面的相关知识，并运用到建筑节能设计中。

一、课程介绍

南京大学建筑与城市规划学院的前身是2000年建立的南京大学建筑研究所，以前主要从事研究生教育。从2007年起，开始招收建筑学本科学生。在2010年与南京大学城市规划系合并，改称为南京大学建筑与城市规划学院。南京大学建筑与城市规划学院建筑系一直积极探索和尝试建筑学的教育改革，致力于培养具有坚实基础理论和宽广专业知识的高层次实用型人才。南京大学建筑学本科教育建构的是与国际建筑教育界接轨的4+2模式。即充分利用南京大学基础学科的优势，第一年进行通识教育，后三年进行建筑学专业基础培训。本科毕业后，对建筑学有兴趣及能力较好的学生可以进入研究生教育，进行建筑学专业高级培养。相较于我国其他五年制建筑学教育而言，这种模式的优点就是南京大学建筑学本科学生具有较坚实宽广的基础知识，但缺点就是建筑专业基础理论课不得不压缩在较短的时间内，这给教师教学与学生学习都带来较大的压力。

笔者教授的是建筑学本科三年级下半学期的建筑技术（二）——建筑声、光、热（课程号：291170）和建筑技术（三）——建筑设备(课程号：291180)。每门建筑技术课程的跨度是16周，共38个课时。建筑技术（二）即是传统的建筑物理课程。本门课程目标是使学生能熟悉掌握建筑热工学、建筑光学、建筑声学中的基本概念和基本原理；了解建筑的热环境、光环境、光环境的质量评价方法，以及相关的国家标准；具备在数字技术方法等相关资料的帮助下完成一定的建筑节能设计的能力。

建筑技术（三）——建筑设备课程介绍了建筑给水、排水系统、采暖通风与空气调节系统、电气工程的基本理论、基本知识和基本技能。本门课程目标是：让学生能熟练地阅读水电、暖通工程图；了解水电及消防的设计、施工规范；具备燃气供应、安全用电及建筑防火、防雷的初步知识；能够根据相关资料来评估与解决与建筑设备相关的建筑、环境问题；了解各类建筑设备系统的特性、系统要求、系统布置，以及与其建筑物的相互关系，并能灵活地运用到建筑设计中。

二、教学探索

我国传统的建筑学院在这两门课程教学上偏重基础理论教学，但主要问题是内容涵盖面较窄，一些的技术理论和知识相对滞后，基础理论点过窄，但与建筑实践联系较少。与我国在建筑设计教学上的不断改革和创新不同，建筑技术课程教学在我国建筑学院、系里基本上还保留着传统的填鸭式教学，即一门课、一本书、一次考试、一次性归还给教师的模式。虽然近几年来我国建筑学界也一直在尝试改进建筑技术课程，但总体上变动不大。因此，笔者希望能从课程内容、形式和作业布置上，采用引领式、开放式、双向式教学，主要包括以下几个方面的探索：

1. 课程在内容上除满足我国建筑学专业指导委员会所规定的基本要求外，还参考了新加坡和中国香港地区大学以及美国麻省理工大学建筑系的建筑技术课程内容，增添许多较新的知识点，如美国LEED（Leadership in Energy and Environmental Design）的标准介绍，以及从单体到小区整体环境规划方面的主动和被动节能设计理念。此外，笔者加入了大量的经典和最新的建筑实例的介绍，尽量穿插在基础理论部分。建筑实例的介绍不仅吸引学生的学习兴趣，而且还能开阔眼界，使他们了解国内外最新建筑节能的发展趋势。虽然课程也采用了我国建筑学专业指导委员会所建议的普通高等教育土建学科专业"十一五"规划教材[2][3]，但为配合课程内容，每次课笔者都会给出附加的大量中英文参考阅读书目，列出相关的国内外网站，要求学生能够做到课前和课后阅读，提高学生的自我学习和研究的习惯，让其从整体上拓宽和提高建筑技术方面的知识。

2. 由于计算机能耗模拟技术在辅助建筑节能设计领域开始发挥重要的作用，引入计算机能耗模拟软件教学势在必行[4]。在能耗软件学习上，笔者选择介绍Ecotect和Airpack两款软件，以学习Ecotect为主，Airpark作为辅助风模拟分析。选择Ecotect的主要原因是：(1)该软件是一款可视化程度较高的计算机能耗模拟软件，能够在热环境模拟、日照分析、造价分析、声学分析、光环境模拟等方面帮助建筑师进行优化设计，直观可视的计算结果，很受我们建筑设计院的建筑设计人员推崇。尤其其全面的能耗模拟和分析功能，能够配合建筑物理中的声、光、热的基础知识的学习[5]。(2)目前很多国外大学的建筑系，都将Ecotect的学习纳入他们必修的课程内，我国一些大学的建筑学专业也开始将该软件列为建筑物理课程的教学中[6]。(3)Autodesk公司允许免费使用，并提供相应的辅助教材和教程。因为要与建筑设计课相配合，能耗模拟软件的学习穿插在两门技术课程中，共占用8个课时。

3. 教学理念上强调理论与实践相结合。如在建筑物理教学上，要求学生走出教室，对实际建筑的声、光、热进行实地测量；对照我国现行的标准和规范进行数据分析与评估，并通过计算机模拟分析进行建筑节能设计改造。对于建筑设备课程，则更是偏重教授实践知识，相较于单纯的介绍建筑设备的原理和要求，笔者通过建筑实例来说明建筑设备与建筑设计之间的关系。实地学习是教授建筑设备最好的方法之一，带领学生带着设备图纸到现场进行观察、实地测量，通过对比学习来真正认知建筑设备与建筑设计之间的关系。

4. 在作业安排上，尽量减少死记硬背的考试，增加课程报告和课程设计内容，让建筑技术课程的学习，从理论知识与建筑设计有机联系起来。南京大学建筑三年级下半学期的建筑设计课题是住宅和小区设计训练。建筑物理的课程报告是以住宅建筑为主。基本上包括两部分：一是小组作业，主要是选择一个真实的住宅单元进行现场调研，实地测量声、光、热环境，并要求运用所学的计算机模拟软件做节能改造设计的定量分析。该训练不仅可以让学生巩固建筑物理的声、光、热基础知识，也对建筑实际的物理环境有一个基本的认识。第二部分单人作业则是以建筑课程设计为依托，对自己设计的住宅单元进行节能分析和节能改造设计；建筑设备课程作业也与建筑物理相类似，首先进行实地走访和认知学习，分析和总结设备与建筑设计存在的问题，个人作业是针对自己的住宅设计，进行水、电设计。

5. 在教学形式上，笔者除自己讲课外，也采用播放录像、实地考察和学生讲演方式来活跃课堂气氛，并经常鼓励和组织学生进行课堂讨论，培养学生自主思考能力。根据课程的内容笔者还会聘请一些专家来做专题讲座，以丰富学生的知识面，扩大学生的视野。笔者发现"走进来和走出去"课堂教学方法，对学生学习知识有很大益处。如为了解建筑设计和设备之间的整合关系，笔者邀请设计院的工程技术人员走进教室，与学生面对面畅谈他们自己的实践经验，同时鼓励学生走出教室，自己去观察、学习真正的建筑设备，亲身体会设备的要求、尺度以及与建筑设计之间的关系。

总而言之，建筑技术课程改革与革新的最终目的还是要回归设计，即如何将与技

AN INNOVATION EXPERIMENT OF BUILDING TECHNOLOGY TEACHING IN AN ARCHITECTURE SCHOOL • WU Wei

术相关的理论和知识更好地服务于建筑设计,这也是国内外建筑学专业在教授建筑技术课程上所面临的主要问题。笔者以上的探索和尝试,强调的都是如何将死的理论灵活地运用到活的设计上。这不仅仅需要二者在课程内容、时间安排和作业设计上的整合,更需要二者的指导教师在整体教育思想上有一定的共识,感谢教授建筑设计课的华晓宁老师,我们都为建筑技术的教改做出了不少探索和尝试。

三、主要问题和经验教训

因为已经将传统的建筑物理与设备课程压缩到每门36个学时,加之笔者又增添了许多新的教学内容,所以只有在课堂教学内容上弱化一些基础理论知识,尽量提纲挈领,强调培养学生的自我学习、实践认知以及一些基本的研究能力。旨在帮助学生认识建筑技术的必要性和重要性,拓宽和提高建筑技术方面的知识,以便更好与建筑设计相配合。然而,在教学探索中,笔者也遇到了不少问题:

1. 由于我国学生更习惯于被动式、封闭式、单向式教学模式,经常是老师教多少,学生学多少,即使到了大学阶段也一般缺乏自主学习能力。以课程阅读为例,很多同学普遍没有课前阅读的习惯,当课程进度很快,每节内容涵盖很大时,会有个别学生表示难以跟上课程进度。其次,一些同学经常出现不知道读什么、怎么读和什么时候读等问题。特别是我国学生从小就习惯于一门课一本书这种填鸭式的学习模式后,当被要求进行大量资料的泛读,以及从大量的参考书目中自主选择阅读时,学生们则显得有点无所适从,力不从心了。他们显然不太习惯根据自己和课程的需要筛选资料,分门别类地进行泛读和细读。尤其是直接让他们阅读英文原著时,大部分学生还是表示有一定困难,这可能与我国英文教育以应试培养有关。笔者也曾尝试组成读书小组,以期能互帮互学,但对于这种国外学校学生普遍采用的互助式学习方式,我国学生还不太习惯和适应,基本上还是各自为政。

2. 引入计算机能耗模拟技术的学习,不仅可以帮助学生更扎实地掌握基础理论知识,也能更好地开展建筑节能设计。但计算机能耗模拟软件与传统的理论教学不同,笔者在Ecotect教学中发现很多问题:(1)在建模阶段,最容易出现的问题是混淆计算机辅助建筑设计软件(CAAD)与建筑能耗模拟软件在使用上的异同,如模型简化和模型精准性问题。(2)参数设定时的主要问题是材质设定。(3)盲目追求各种模拟分析图表,但对模拟结果不求甚解。(4)对计算机能耗模拟软件产生认知错误。如对能耗模拟结果产生盲目认同,甚至认为软件可以代替设计者做节能设计,当发现不可能自动生成节能设计时,又对软件的作用产生质疑。

3. 在课堂上,笔者为鼓励学生自主思考能力和活跃课堂气氛,会常常鼓励学生回答或提出问题,并组织学生进行分组讨论。但令人失望的是,同学们很少能自愿回答问题,更别说自主提出问题,在课堂讨论时,也不愿发表意见,因此很难形成真正意义上的课堂讨论。

4. 建筑技术与建筑设计课程在互相配合上还是需要进一步探索。虽然在每学期的开始,笔者都与建筑设计课程的指导主教师进行推敲,尽量跟建筑设计课在内容、时间安排、作业上保持一致。但由于课程性质的不同,总会出现一定的问题。例如建筑技术的一部分课程作业是依托在课程设计上,因此希望该作业能够成为课程设计的一部分。但如何安排提交课程作业的时间就是一个大问题。如提交时间太靠近设计答辩时间,会影响最终课程设计的出图和答辩。但如距设计答辩时间太远,学生的设计户型和单体确定不了,他们的水、电设计以及建筑节能设计无法真正反映到课程设计上。即使笔者与设计课教师尝试将二者的设计合二为一,但学生往往顾此失彼,大部分仅专注于整体设计,而忽视后面的水、电以及单体的节能设计。此外,还有一个令人值得注意的地方,就是当学生在利用计算机能耗模拟软件进行环境辅助设计时,笔者发现学生会被软件各种各样的分析功能所迷惑,简单地将各种模拟分析结果放在自己的建筑设计中,认为这就是建筑节能设计了。

尽管出现了不少问题,但值得欣慰的是,绝大部分同学还是非常刻苦认真地学习,并十分愿意配合和响应老师在教学上的不同尝试。当课程结束后,笔者对学生做了问卷调查和个别访谈,几乎所有学生都认为建筑技术课程很重要,他们学到了很多

有用的东西,对开阔他们的设计思路很有益处,认为学到的知识对今后发展和提高有一定帮助。同时,他们都觉得老师的教学方法较新颖有趣,能够帮助他们掌握和思考所学知识。同学们反映的主要问题集中在课堂信息量大,节奏过快,在完全掌握上有一定难度,加之大量的课外阅读和课程作业,使他们感觉负担过重。

四、结语

随着我国面临着越来越严重的环境和能源危机,在建筑设计和施工中实施可持续发展战略已经不是一种选择,而是势在必行。因此作为建筑节能设计基础的建筑技术教育,也日益受到建筑学界的日益重视,这无疑为传统的建筑技术教学提出了新的机遇和挑战。可以预见,建筑技术课程如不积极创新和改革传统模式,将无法适应当今世界的发展。

笔者希望通过对这两门建筑技术课程的改革和创新探索,对国内同行有抛砖引玉的作用。笔者认为西方先进国家的本科教学培养模式,即采用引导式、开放式、双向式的教学,可以更好地培养学生自主学习能力。因为大学期间的建筑技术课程所能教授的理论和知识是极为有限的,只有让学生认识到建筑技术的重要性,以及正确的自主学习方法,才是保证学生学到真正知识的根本。根据这两门课程结束后的回访,很多同学都表示建筑技术影响和开阔了他们的设计理念,所学到的学习方法对学习其他课程也有一定帮助,笔者认为这已经达到了预期的改革教学目标。

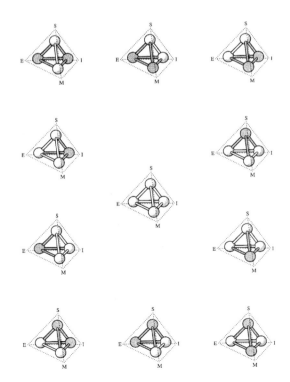

In 1969, in the introduction of his book *The Architecture of the Well-tempered Environment*, Reyner Banham appealed for an "Unwarranted Apology" to classify his book under "history and theory of architecture" instead of under "an introduction of technology"[1]. However, forty-three years later, his book is still under the same section. Sadly, the division between architectural design and technology, as prophesied by Banham, has entrenched both in practices and in design studios. Undeniably, the building technology strong affected modern architectural design, and its related courses such as "Environment Sciences" and "Building Service Systems" are important part of architectural education today. However, it is well known that architectural students would like to ignore these courses, in particular in China. How to better integrate building technology knowledge into architectural design is a hot topic in both architecture education and design practice.

In the last decade, the sustainable design has played an important role in architectural field. And building energy conservation has very significance in China. It is necessary for architects to master more technology-related knowledge. This is not only a new opportunity but also a challenge for technology education in architectural school. Therefore, the technology courses in architecture schools need to reform its context and education methodology. Beside the traditional knowledge, the architectural students need to know some cutting-edge technology, and the most important thing is to integrate technology into design process.

This paper introduces an experience of teaching two building technology courses to senior undergraduate students in School of Architecture and Urban Planning, Nanjing University, P. R. China. The author analyzed and summarized the major problems during the experimental teaching reform. The motivation of this teaching reform is to develop strategies to synchronize knowledge of building technology into architectural design, in particularly in architecture schools in China.

I. Course introduction

Graduate school of Architecture, Nanjing University was established in 2000, the school had placed a heavy emphasis on graduation education for many years. It started to enrolled undergraduate students in 2007. It was merged with the Department of Urban Planning, Nanjing University two years ago and the two became School of Architecture and Urban Planning. The Department of Architecture dedicated for exploring architectural education and teaching reform from its start day. Its major goal is to build multi-perspective and diversified of talent development.

The undergraduate education in Department of Architecture in Nanjing University is so called "4+2" mode which is four years' undergraduates plus two years' postgraduates. This kind of model is one of popular architectural education models in the west world. The first of the four years is liberal education which is the advantages of Nanjing University. the last three years is professional architectural education . The undergraduate students who are really interested in architecture and have good academic achievements could continue their post-graduated studies. Compared to other five-year architectural education in China, the advantage of this mode is that the undergraduate architectural students from Nanjing University have a solid and broad foundation of general education; the shortage is that the traditional undergraduate architecture education have to be compressed into a shorter time period. This brings greater pressure to both the teachers' teaching and the students' learning.

What the author teach is building technology II — acoustics, lighting, and heating environments (course ID: 291170) and building technology III — building service systems (course ID: 291180) for the second half semester of architectural undergraduate in grade 3. The teaching period of each building technology course is 16 weeks, a total of 38 class hours. Building technology II introduces the basic concepts and fundamental theory of the building thermal, lighting and acoustical environments. It then focuses on the quantitive evaluation method, including use of instrumentation and thoughtful analysis of data. Computer simulation, including modelling and interpretation of results is applied to study passive energy design

Building technologyIII–building service system course introduces the basic principle of building service systems such as water supply and drainage systems, HVAC systems, and electrical engineering. This course objectives are: (1) A good understanding of the major building services systems and their integration and coordination into the architecture; (2) An understanding of the working relationship between the architects and the building services engineers' professor practice;

(3) An ability to assess and solve some particular problems, which relate to technical performance, economics, energy usage and environmental effort.

II. Experimental study

The teaching tradition of these two courses in China's schools of architecture emphasizes the basic theory, of which the main problem is that some of the technical knowledge with narrower content coverage is lagging relatively behind, the basic theory is too abstruse, and it is not sufficiently linked to the architectural design. Different from the continuous reform and innovation in the teaching of architectural design, the teaching of building technology courses in China's architectural schools and departments essentially retains the traditional cramming system, i.e. the mode of one course, one main text book, one examination, and a all-in-one return what they

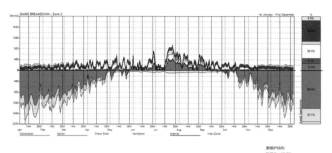
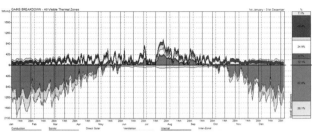

learned to the teacher. Although architectural educators in China have been trying to improve these technology courses in recent years, no significant change has been made by and large. Therefore, the author hopes to carry out an experiment which uses an open, and two-way teaching method with respect to the course's content, form, and homework arrangements, it mainly includes the following aspects:

1. Beside the requirements stipulated by the Committee of architectural education in China, the course also adds some new knowledge and information based on references of some universities in Asia, such as the National University of Singapore and the Chinese University of Hong Kong, as well as some well-known western universities in the Department of Architecture of MIT. The cutting-edge technology and information, such as advanced building energy simulation, Evaluation Standard for Green Building in China and community green planning etc. are introduced. Some state-of-the-art green design projects also, are given. These real design projects, not only can attract the students' learning interest, but also broaden their design knowledge related to technology. Although the courses still use the main text books which are recommended by the Committee of Architectural Education in China[2][3], the author also gave a reference reading list at each class, which included the related books and even websites both in English and Chinese. The required pre-course and after-course readings could help the students develop self-learning ability, and broaden and improve their overall knowledge in the field of building technology.

2. As building energy simulation techniques start to play an important role in the green building design, introduction of teaching with building simulation software is imperative [4]. Two building simulation software, Ecotect and Airpark are introduced during the courses, Ecotect is main energy analysis software, and Airpark serving as the CFD related analysis. The reasons that the author chose Ecotect are that: (1) Ecotect is a high visualization software, and it can help the architects do thermal, lighting and acoustical environment simulations, as well as sun shading and cost-efficient analysis,etc. The intuitive and visualized simulation results of Ecotect are welcomed by designers. In particular, the different kind of building environmental analysis could help the students understand the theory of building technology better [5] .(2) Ecotect has been included either in their technology related courses or design studio in many Western universities, some architecture department/school in China also start to introduce Ecotectin building technology courses [6]. (3) The Autodesk Company agrees to give the software for free and provided appropriate support materials and tutorials. To be compatible with the design studio, the software learning is interspersed into Building technology II&III, which is a total of eight lessons.

3. The teaching philosophy emphasizes an integration of theory, field trip and design studio. For example in the teaching of theory of building science, students are required to go out of the classroom to do on-site measurements as well as study how to do data analysis and building performance based on China's current standards and rules. They are also asked to do energy efficiency design through the computer simulation software. The course of Building Service Systems emphasizes on how to integrate building services into architectural design, instead of only introduction of building service systems, the author gave many design examples to illustrate different integration methods. The author believes that field study lets architectural students directly see, touch and even measure the building service systems in a real building, and consequently have a better understanding of the integration.

4. The author found the course projects which emphasize on on-site measurement and total building performances have better effects of learning than the final examination. The design studio parallel with the technology courses is design projects of residential buildings and community. Therefore, the course project of the building science asked the students to choose a residential building, to study its thermal, lighting and acoustic environments. The course project includes two parts: the first part is group work which is on-site measurements and environment analysis based on measured data for the existing residential building. The second part is individual works which do a energy-saving reform for this existing residential building and use building simulation software to exam their design, on-site measurements and existing environment analysis, and individual report, This course project aims to better understand of the basic concepts and fundamental theory of the building thermal, lighting and acoustical environment and to apply computer simulation to do green building design. Similar to the build technology II, the assignments of technology III is to study service systems for a small public building, the students was asked to find and analyze any problems that building systems does not well integrate with spatial design.

5. In order to improve architectural students' interests for learning building technology courses, different teaching methods such as play videos, field visits, and guest lectures are arranged. Class discussions which could help students develop independent study ability are strong encouraged. The guest lectures which the author invited experts and engineers in design company, give architectural students a chance to gave a face-to-face talk. These experts' real practical experiences enrich the students' knowledge and expand their horizons. Also, the author encourages the students to go out of the classroom to observe and study the buildings service

systems that they learn in the classroom. The author found that so called "coming in and going out" teaching method can attract students' interests and learning efficiency. All in all, the ultimate purpose of these experiments is toward a better design, that is, how to let the technology-related theories and knowledge better serve the architectural design. This is an old problem facing to all architecture schools/departments at home and abroad. The above exploration and attempts emphasizes how to flexibly use the pure theory in the design. This not only requires the integration of the technology courses and the design studio in the aspects of teaching content, timing, and course projects, but also requires two lecturers have a certain consensus in the overall education ideologically. Great thanks should be given to professor Hua Xiaoning who is teaching leader of the design studio at the third year, his cooperation made this experiment to be possible and meaningful.

III. Discussion

In general, a total 36 teaching hours is not enough for traditional technology courses in architecture schools/departments in China. The author also added many new content for these two technology courses. As a result, the author has to reduce some less important content, in particular for some knowledge that the students could easily study from their main text book, and put emphasis on introduction of cutting-edge technology and real design projects. The key is to let architectural students to know how the building technology is important to their design and how to use this knowledge to improve their environment design. However, this experiment also encountered many problems: the necessity and importance of building technology.

1. Since most Chinese students are more accustomed to the passive, closed, one-way teaching mode, the students learn as much as the teachers taught, so they are usually do not get used to self-learning even after entering college stage. Taking course reading as an example, many students generally do not have the habit of reading before class; when the course progress is faster and each lesson covers a lot of content, some students would feel some difficulties to keep up with the course process. Secondly, some students often do not know what is intensive and extensive to read. Especially, our students grew up accustomed to one main text book for one course (cramming system), when given a extensive reading list and asked to pick up an intensive reading list by themselves, the students seem a bit confused and powerless. They are obviously not accustomed to extensive reading and peruse according to their own needs and curriculum needs. Especially when they are required to directly read the original English version, the majority of students expressed certain difficulties. This may be related to China's English education and exam-oriented culture. The author has also suggested the students organize study group for helping each other. However, it seems that most of the Chinese students are not accustomed to this collaborative learning way widely used by students in Western education.

2. The introduction of learning building energy simulation technology can not only help the students grasp the basic theoretical knowledge better, but also better carry out sustainable architectural design. However, teaching building energy simulation software is different from the traditional theoretical teaching, and the author has found many problems in Ecotect teaching: (1) At the modeling stage, the commonest problem is to confuse the differences and similarities in use between the computer-aided architectural design (CAAD) software and building energy simulation software, such as model simplification and model precision problems; (2) The major problem in parameter setting is how to define the material; (3) Since the architectural students have difficulty to understand various simulation analysis and charts, they have trouble applying these simulation into their design, simulation blind; (4) Misunderstanding of building energy simulation tools, such as blind acceptance of simulation results, and even believing the software can replace the designer to make energy-saving designs. When the students found out that the simulation tools could not automatically generate the energy-saving design, they began to question the software.

3. At the class, to the students' independent thinking are strong praised. For enliven the classroom atmosphere, the author often encourages the students to answer questions or ask questions, and organize them to have group discussions. However, it is disappointing that a few of the students would like to answer questions, So, the author had no choice to pick up the students to answer the questions., And the students told the authors after the class that answering questions during the lecture does help them pay more attention on the lecture, but they felt quite stressful when they were picked up.

4. Further exploration on the cooperation of building technology course and design studio is still necessary. Although at the beginning of each semester, the author often deliberates with the teaching leader of the design studio to try to be consistent in the content, timing, and projects of the design studio. However, there will always be a certain problem due to the different natures of the course and studio. For example, part of the course projects is based on the design studio's projects, and therefore these course projects could become a part of the design studios' work. However, the date of submission along is a problem. If the submission time is too close to the submission time of the design studios, the students would complain no time to do

their design work. But if too far from the submission time of the design studio, they cannot finish their sketch design, and their design for building service systems and energy efficiency design can not be truly reflected in their projects of the design studio. Even if the design teachers and the author have tried to combine them into one design project, the most students are often only focus on the spatial design, while ignoring the subsequent water supply and electricity design, as well as its energy-saving design. And a couple of other things to keep in mind, the architectural students would like to put all kinds of simulation images without a fully understanding. They thought more simulation results means good energy saving design.

Despite a lot of problems, the vast majorities of students in architecture department are hard-working students, and are very willing to cooperate with and respond to the teachers' experiments. For better understanding the experiment results, a questionnaire survey and a students' interview were carried out by the teacher assistant after the course finished., Almost all students thought that building technology is very important for their design. They learned a lot of useful knowledge that were very beneficial to broaden their design ideas, and they believed that the learned knowledge would be helpful for their future development. Meanwhile, they felt that the teacher's teaching method was relatively new and interesting, and would help them master the knowledge. The main problems proposed by the students include the large amount of teaching material each class. Sometimes, they have trouble to following the author's teaching without pre-course reading. And course projects plus the design studio's work make them feel overwhelmed.

IV. Conclusion

As China is facing increasingly serious environmental and energy crises, the strategy of sustainable development in building industries is very important and urgent. Since the knowledge related to the building technology is the foundation of green building design, technology courses in architecture schools/departments started to attract the architectural educators' attention. This is undoubtedly the new opportunities and challenges for the traditional building technology teaching. It expects that if building technology courses in China is still using the traditional teaching content and model, it will be very hard for our tomorrow' architects to catch up with sustainable design.

Through experiments of these two building technology courses, the author hopes to serve as a stimulus to the domestic counterparts. The author believes that the teaching methods in the Western countries, such as encourage students' independent leaning and thinking would help students to master the knowledge. Since the theories and knowledge which can be taught in the university are extremely limited, allowing students to recognize the importance of building technology and the right self-learning method is fundamental to ensure that students have better understanding of the knowledge and apply it in their design projects. Based upon the questioner survey and interview after the course, most students said that the knowledge learned from these two courses are very useful in their design, and open their design's view. And many students stated that they liked this teaching style, and this kind of teaching helps them better master the knowledge, the author believes it has reached the expected objectives of the experiments.

References

[1] Reyner Banham. Architecture of the Well-tempered Environment [M]. [S.L.]: Architectural Press, 1969.
[2] Liu Jiaping (Editor). Building Physics (Chinese) [M]. Beijing: China Architecture & Building Press, 2009.
[3] Li Xiangping, Yan Zengfeng (Editor). Building Services Systems (Chinese) [M]. Beijing: China Architecture & Building Press, 2008.
[4] Massimo Palme. What Architects Want? Between BIM And Simulation Tools: An Experience Teaching Ecotect [A]. Proceedings of Building Simulation 2011: 12th Conference of International Building Performance Simulation Association, Sydney, 2011: 2164-2169.
[5] Li Kun, Yu Zhuang. Design and Simulative Evaluation of Architectural Physical Environment with Ecotect[J]. CADDM, 2006, 16(2): 44-50.
[6] Yan Jun, Zhao Neng, Liang Zhiyao. Application Ecotect in Architecture Design[J] .Journal of Architectural Education in Institutions of Higher Learning, 2009, 18(3): 140-144.

CAAD基础教学新思路・童滋雨 刘铨

一、CAAD基础教学定位

CAAD已经成为建筑设计不可或缺的工具，尤其体现在建筑制图方面。而在建筑教学体系中，对CAAD课程的定位仍然存在着多种不同的意见，其差异性体现在教学起始的阶段、课时以及内容等方方面面。从这一点上来说，CAAD教学尚未形成一种标准的模式，存在着很大的讨论空间。

CAAD基础教学是指建筑学专业学生初步接触CAAD所学习的课程。在早期的CAAD基础教学中，主要内容是教授各种计算机绘图软件的使用，这使得该课程更像是一门软件学习课，缺乏鲜明的建筑学特征。另一方面，本科阶段的CAAD教学大多设置在掌握了基本的建筑制图知识和经过基本的设计训练之后，作为一种必须掌握的基本工具，这样的设置也偏晚了。因此，我们试图对CAAD基础教学进行重新定位，不但要加强该课程与建筑设计之间的关系，而且要将其作为必备工具尽早应用于设计课程。

在新的定位要求下，CAAD基础课程将作为学科通识课程，在学生接触建筑学专业知识之初就与其他专业课程同步进行教授。我们将建筑制图知识融入CAAD基础教学中，将其中心任务从以往的重点讲述怎样使用绘图软件，向真正的辅助设计转变——即如何借助计算机工具画出符合建筑设计要求的图纸，强调计算机绘图在推动设计中的作用以及图纸对设计意图的最终表达效果。同时，课程进度设置与同时期的建筑设计基础课程同步，所有案例和课后练习直接来自设计课程的内容。总的来说，新的CAAD基础课程融合了建筑制图、空间认知、建筑表现、图面排版等多方面的内容，帮助学生尽快掌握建筑设计的基础工具。

二、CAAD基础教学目标

CAAD基础课程在南京大学建筑学教育的体系中被划分为学科课程通识的范畴，时间被安排在本科二年级上学期，与"建筑设计基础"课程同步进行。在该教学体系中，二年级是学生真正接触建筑学知识的开始，因此，CAAD课程不但要承担通常的CAD类软件使用的教学工作，让学生会用计算机进行建筑表达；而且要和设计课程一起，承担起建筑制图知识的教学，让学生掌握正确的、各种类型的建筑表达方法。在这样的要求下，本科阶段CAAD基础课程目标被初步设定为：掌握建筑制图的基础知识，并能应用CAD类软件进行规范的建筑表达，包括二维制图和三维模型等。

在明确的教学目标的指引下，学生对基本建筑表达的语言得到了比较全面的掌握。然而教学实践成果的反馈也表明，学生对设计内容的表现仍显得单调、平淡。尤其是全部采用计算机出图，由于屏幕显示和实际的差异，在图纸表达和预想效果之间有着相当大的差距。因此，我们对CAAD课程的目标和内容进行了反思，在原来课程的基础上，进一步强调了最终的图纸表现力，从学会用计算机进行正确的建筑表达上升到学会用计算机画出具有更强的建筑表现力的建筑图，为将来能够更好地表达设计意图、设计过程和空间效果积累知识。

根据这些反馈和反思，本科阶段CAAD基础课程的目标被修正为：掌握建筑制图的基础知识，能应用CAD类软件进行建筑图的绘制，并加强纸面效果的表达能力。

三、CAAD基础教学内容

基于CAAD基础课程的教学目标，我们在CAAD课程中设置了两条知识脉络——建筑设计知识脉络和建筑制图知识脉络（图1）。这两条知识脉络的设置与同时开始的主干课程"建筑设计基础"密切相关。

在建筑设计知识脉络方面，CAAD基础课程的全部案例和练习都来源于"建筑设计基础"课程的教学内容。"建筑设计基础"课程以对建筑立面认知测绘为起点，逐渐扩展到建筑空间、建筑构造、建筑形体和建筑环境。而CAAD课程同样以建筑立面作为起点，介绍其在建筑制图中的特点，并利用计算机软件对其进行绘制，然后扩展到平面图、剖面图、轴测图和三维模型。

在建筑制图知识脉络方面，CAAD基础课程首先以正投影作图为基础介绍建筑立面、平面、剖面的制图原理和方法，进而结合墨线图的表现讲解在二维图基础上阴影的求解方法、轴测图的画法等，最后才是三维模型的透视图表现特点和技巧。

从这两条知识脉络可以看出，尽管是CAAD基础课程，但计算机软件本身并不是重点的讲解内容，而只是作为一种制图工具，工具的掌握是为了更好地表达设计内容。在课程中我们没有专门针对某个软件进行细致入微的全面介绍，而是结合教学的建筑图纸内容，如墨线图、单色图、效果图等，根据其需要，讲解具有较强针对性的某种计算机软件的某一部分功能。这样的教学设置不但可以与建筑设计课程建立更紧密的联系，而且避免了软件更新过快而带来的教学内容容易过时的问题。在实践中，我们分别选择了AutoCAD的二维制图功能、SketchUp的三维建模功能、Vray的渲染功能、Photoshop的图像处理功能和Indesign的排版功能。

为充分体现CAAD基础教学重视纸面效果表达的目标，本课程要求学生将每一次的课外练习成果都以打印的形式上交，教师再根据其上交成果的具体体效果进行点评，并给出相应的改进建议。在实践中我们发现，学生作业中的主要问题都不是计算机软件的掌握问题，而是建筑制图问题和纸面效果表达问题，这也证明了我们进行教学改革的必要性和合理性。

四、结论

南京大学建筑学本科CAAD基础教学摒弃了以软件为核心的教学方式，逐步从会计算机制图发展到用计算机进行正确的建筑表达，再发展到用计算机进行准确且具有明确意图和表现力的建筑表达，这是对课程核心价值的回归，充分反映了课程的专业性和特殊性。

南京大学建筑学本科CAAD基础教学的内容分别以建筑设计和建筑制图为两条知识脉络同时展开，为以建筑设计为主干的人才培养分类贯通创新模式提供了有力的支撑。

由于新的教学目标和内容完全开展的时间尚短，其中还可能存在许多不足，这些也都有待于进一步的经验积累和反馈，使其可以取得更好的效果。

1. Positioning of basic CAAD teaching

CAAD (computer aided architectural design) has become an indispensable tool for architectural design, especially in the aspect of architectural drawing. While in the architecture teaching system, there are still several different opinions regarding the positioning of CAAD courses, of which the differences are reflected in such aspects as the initial semester, class hours and contents, etc. On this point, CAAD teaching has not yet formed a standard mode, leaving a lot for us to discuss.

Basic CAAD teaching means the courses taken by the architecture students upon their initial contact with CAAD. The earlier basic CAAD teaching focused on teaching the usage of various CAD softwares, making this course more like a software learning course with no distinctive architectural features. On the other hand, the CAAD teaching at the undergraduate stage is mostly set after mastering the basic knowledge of architectural drafting and design. It's a little late for such a necessary tool. So, we are trying to reposition the basic CAAD teaching, not only strengthening the relationship between this course and the architectural design, but also taking it as an essential tool to use it in the design courses as early as possible.

Under the requirements of the new positioning, the CAAD foundation courses will be taught as a general disciplinary course synchronously with other professional courses when the students start to study the knowledge of architecture. We integrate architectural drawing into the basic CAAD teaching, changing its central task from the original teaching how to use CAD software to the real aided teaching, i.e. how to plot satisfactory drawings with the aid of a computer tool, how to understand the significance of computer tools in the design process and how to improve the expression of the design intent. The course schedule is also in line with the basic courses of architectural design of the same period, with all cases and exercises directly coming from the design course. Overall, the new CAAD foundation courses integrate the architectural drawing, spatial cognition, architectural expression, drawing layout and so on to help students master the basic tools of the architectural design as soon as possible.

NEW IDEAS ABOUT BASIC CAAD TEACHING • TONG Ziyu   LIU Quan

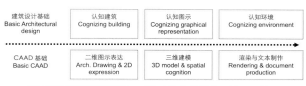

Fig. 1 Basic CAAD teaching workflow

2. Objectives of basic CAAD teaching

In Nanjing University's architectural education system, the CAAD foundation course is scheduled into the general courses, being taught in the first half semester of undergraduate Grade 2, which is in line with the course "Basic Design of Architecture". In this teaching system, the students begin to get in touch with the architectural knowledge in Grade 2. Therefore, the CAAD course will not only include teaching the usage of usual CAD softwares to enable the students to use computer to express the design, but also include teaching the knowledge of architectural drawing to allow students to master various types of expression of design. In view of such request, the objectives of the CAAD foundation course at undergraduate stage are initially set as mastering the basics of architectural drawing and being able to use CAD softwares to carry out normative architectural expression, including the two-dimensional drawings and three-dimensional models.

Under the guidance of the clear teaching objectives, students have a more comprehensive understanding of the basic architectural expression language. However, the feedback on the outcome of the teaching practice also shows that the students' performance over the design content still seems monotonous. Especially due to the differences between screen display and the drawings paper, there is a considerable gap between the final drawing effects and the expected effect. Therefore, we rethink about the objectives and contents of the CAAD course and decide to further stress the expressive effect of the final drawing, evolving form learning how to use the computer to express a building correctly to learning how to use the computer to plot drawings with more architectural expressive effect.

Based on this feedback and rethinking, the objectives of the current CAAD foundation course is modified as better understanding the basics of architectural drawing, being able to draw the architectural design using CAD softwares and improving the print effect expression.

3. Contents of basic CAAD teaching

Based on the objectives of basic CAAD teaching, we have set two knowledge contexts in the CAAD curriculum — building design and architectural drawing (Fig 1). The setup of these two knowledge contexts relates closely to the backbone course "Basic Design of Architecture".

In the context of architectural design knowledge, all the cases and exercises in the CAAD foundation courses are from the teaching contents of "Basic Design of Architecture". "Basic Design of Architecture" starts from the cognition of surveying of the building elevation and gradually extends to the building space, building construction, building form and building environment. While the CAAD course also starts from the building elevation, to introduce its characteristics in the architectural drawing and draw it using computer software, and then extend to the plans, sections, isometric drawings and three-dimensional model.

In the context of architectural drawing knowledge, CAAD foundation courses first introduce the cartographic principle and method for building elevation, plane and section based on the orthographic drawing method combined with the performance of the ink drawing. Then the course explains the method for solving the shadow and isometric drawings on the basis of the two-dimensional drawing and finally, the expression characteristics and skills of the three-dimensional model.

As can be seen from these two knowledge contexts, though it is the CAAD foundation course, the computer software itself is not the focus of teaching, but a basic tool. To master the tools is to better express the design content. We did not specifically give a comprehensive introduction to certain software in the teaching, instead, in combination with the building drawings (such as ink drawings, monochrome drawing, rendering, etc.) in the teaching, we explained certain functions of certain highly-targeted computer software according to their requirements. Such a teaching arrangement can not only establish closer ties with the architectural design courses, but also eliminate the out-dated teaching contents resulting from the fast updating of software. In practice, we selected the 2D drawing function in AutoCAD, the 3D modeling function in SketchUp, the rendering function in VRay, the image processing function in Photoshop and the layout function in Indesign.

In order to fully reflect the objectives of basic CAAD teaching that emphasizes the print effect expression, this course requires the students hand in the printed results of every extracurricular exercise to review. The teacher gives corresponding suggestions for improvement. We found in practice that the main problems in the students' work is not about the mastery of computer software, but about architectural drawing and print effect expression, which proves the necessity and rationality of our teaching reform.

4. Conclusion

The basic CAAD teaching for the architectural undergraduate in Nanjing University has abandoned the old teaching methods that focus on software. It gradually evolved from learning to draw using a computer to expressing building correctly using a computer, and then to perform accurate building expression with clear intentions, which is a regression to the core value of the course and fully reflects its professional and special properties.

The contents of the basic CAAD teaching set two knowledge contexts simultaneously - architectural design and architectural drawing. They provide strong support for the innovative classification through the mode of talent training that focuses on the architectural design.

Since the CAAD foundation course has not been fully carried through for long, there may be many deficiencies. We need to accumulate further experience and feedback in order that a better result could be achieved.

常设设计课程
REGULAR DESIGN COURSES

本科二年级
建筑设计基础 • 刘铨 冷天

课程类型：必修
学时学分：72学时/4学分

建筑设计课程第一年的主要任务是让原本对建筑学一无所知的新生建立起基础性的专业知识架构。其主要内容就是建筑认知和建筑表达。认知是主线，表达是方法。认知成果需通过表达方式得以检验，而表达效果和认知成果直接对应。

教案的基本架构是在重新认识建筑基础知识的前提下，将认知与表达作为这门课的教学主线，依照循序渐进的原则，分四个阶段设置了不同的教学任务，每个阶段有其特定的认知对象和认知方法，同时每个阶段的训练都建立在之前一个阶段学习要点的基础上，力图更好地使学生通过认知的过程从一个外行逐步进入专业领域，并为后续的学习打下宽阔和扎实的基础。

其中第一学期的建筑设计基础课包含了前三个阶段的教学任务，第四阶段的教学安排在第二学期的建筑设计（一）课程之中。

建筑的形象对于新生虽不陌生，但"专业地"看待和表述它就是新知识。因此第一个阶段的认知对象就被设定为学生身边经常看到、接触到的建筑，让学生用已有的建筑体验，来帮助"专业地理解"建筑；在已知的表达方式基础上，学会"专业地表达"建筑，让学生从学习之初就在形象与抽象间建立思维上的关联。

在教学的初始阶段，学生首先通过理解"投形"的概念来了解三维的真实建筑是如何被二维平立剖面图所描绘的，其次通过理解图纸比例的概念来了解不同的图纸传达的不同层次的建筑信息。

认知建筑
Building

徒手平立剖面图绘制
Drawings by Hand

建筑立面测绘
Elevation Drawing

建筑平面与剖面测绘
Plan and Section Drawing

窗构造测绘
Detail Drawing

The first year is the initial in the academic education of architectural design. How to make the new architectural students set up the professional knowledge system in Chinese education practice is the fundamental task.

Based on review of the basic knowledge of Architecture, we take cognition and representation as the major ideas and set the program into four step-by-step sections. Specific cognitive objects and cognitive methods are given in different sections and the teaching program of each section is based on the knowledge of previous sections. Our program tries to set a wide and well-knit background for the subsequent design course.

"Basic Design of Architecture" consists first three sections in the 1st semester. The last section takes place in the 2nd semester as "Architectural Design 1".

It is new knowledge to read and describe the buildings professionally for the fresh students. The first phase is to make the students use professional drawings to understand and record the buildings which they face everyday, to combine the concrete figure and abstract drawing: firstly, how to use the plan, section, elevation to describe a building; then, how to use different scales to express specific information.

Undergraduate Program 2nd Year
BASIC DESIGN OF ARCHITECTURE • LIU Quan, LENG Tian

Type: Required Course
Study Period and Credits: 72 hours/4 credits

认知图示的教学沿用了第一阶段的知识，将图示作为认知的对象，学生通过阅读图纸来制作相应比例的实体模型，在思维过程中完成一次认知上的反馈。能按图示进行操作、还原三维建筑空间也是检验上一阶段学习效果的最佳办法。学习的关键不在于认识图示，而在于能否通过阅读图纸来感知相应的实体与空间。

上一阶段从具象的实体到抽象的图的过程在这一阶段被反转，学生通过阅读不同内容、比例的二维的平立剖面图，制作相应比例的实体模型，在思维过程中完成二维到三维的链接和转换。

本阶段认知对象的尺度扩展到城市层面。在全球化的今天，中国更加高速地进入了城市化的高潮时期。作为建筑物赖以生存的基础，城市空间直接影响到建筑的组织策略、形式策略以及建造策略。教案将城市物质肌理形态问题作为城市空间认知的基础，其内容学生也更易把握。

学生通过记录人眼视角的摄影照片、城市地图、SketchUp建模与透视场景模拟、Photoshop制作的城市分析图、PowerPoint的城市调研分析报告，极大地丰富了自身的空间认知与表达手段，更好地促进了对城市空间的认知。

认知图示 Drawing

认知环境 Environment

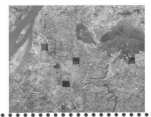

手工实体模型制作 Physical Model

计算机绘图与建模 Drawings and Models by Computer

建筑模型 Building Model

墙身构造模型 Detail Model

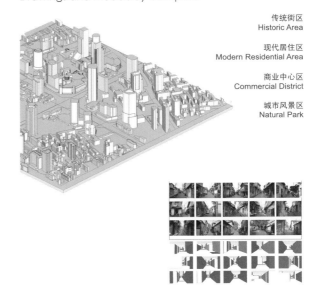

传统街区 Historic Area

现代居住区 Modern Residential Area

商业中心区 Commercial District

城市风景区 Natural Park

In phase 2, the perception process is reversed: students read the architectural drawings to make the physical model with specific scale. The process accentuates the relationship of drawing and building, as well as examines the learning effect of former study. The key point is that the capacity of transformation between 2D and 3D, abstract and concrete in students' mind.

In the globalization world, the urban environment is so important to deeply influence the architectural design strategy of organization, formation and construction. Using photographs, city maps, SketchUp model, and some software tools such as Photoshop and PowerPoint, the students try to understand the urban form through the drawings of street analysis, plot and building analysis and topography analysis.

本科二年级

建筑设计（一）小型公共建筑设计 • 刘铨 冷天

课程类型：必修
学时学分：108学时/4学分

建筑设计（一）完成的是建筑系整个二年级专业主干课程计划第四阶段"认知设计"的训练，课程设置的目的是让学生通过综合运用前几个阶段的建筑知识和表达工具来亲身体验设计的过程。

建筑设计的目的是解决人们对建筑的需求问题。因此，在设计教学中首先需要引导学生认识与发现基本的建筑问题。就建筑设计基础教学而言，这些基本问题就是：功能与空间、场地与环境、材料与建造。初学者由单项问题入手，能更好地理解设计的步骤。学生通过三个练习与一个完整的小型公共建筑设计对前三个阶段学习的知识进行强化与综合，同时逐步认知设计的基本问题与一般过程。

认知设计
Design

设计练习1：先例分析
Case Study

综合的建筑表达与运用
Comprehensive Expression

Architectural Design 1 is the last phase of the whole design course of the 2nd year in the undergraduate program. The purpose is to consolidate the former knowledge and percept the basic design issues and common process through three design exercises and one design project. Students need to use all the architectural knowledge and drawing tools to resolve some basic design issues: program vs. space; site vs. environment; material vs. construction. As an easy way, the students can start with one single issue to better understand how the problem to be resolved in the design procedure.

Exercise Phase 1: Case Study (3 weeks)
Make the physical model based on the plan, section and elevation drawings of the case, and redraw the furniture configuration and analysis of space, function and transportation as the later exercise.

Exercise Phase 2: Program and Space (3 weeks)
Based on the historic area, a traditional housing plot is chosen as the site to design a small shop with some simple living spaces. The exercise focuses on the relationship of function and space as well as the basic structure knowledge.

Exercise Phase 3: Form and Construction (3 weeks)
Based on the modern urban street space, a shop façade is chosen to be renewed. This exercise focuses on the relationship of form, function and construction.

设计练习1：先例分析（3周）
选择若干优秀建筑案例的设计图纸，将其转换为实体建筑模型，同时通过平面图重绘与家具布置、轴测分析图，学习案例的空间构成方式与功能流线安排，作为后续设计练习的铺垫。

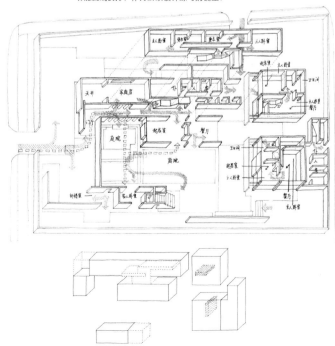

Architecture Program of SAUP, NJU 2011—2012

Undergraduate Program 2nd Year
ARCHITECTURAL DESIGN 1: SMALL PUBLIC BUILDING • LIU Quan, LENG Tian

Type: Required Course
Study Period and Credits: 108 hours/4 credits

设计练习2：功能与空间
Program & Space: Shop Design in Historic Area

Form & Construction: Facade Design for a Shop
设计练习3：形式与构造

设计练习2：功能与空间（3周）
基于城市传统街区环境，选择一个传统建筑所占的地块作为场地，设计自营式商铺，主要关注如何根据建筑功能、流线需要创造建筑空间及学习简单的建筑结构常识。

设计练习3：形式与构造（3周）
基于现代城市街道空间，进行立面设计，主要关注建筑立面的形式及其需要解决的功能问题、建筑材料及其建造问题。

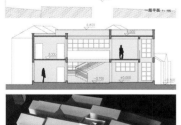
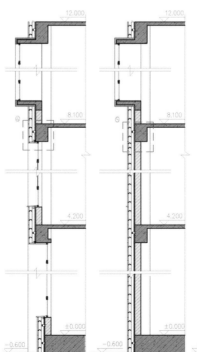
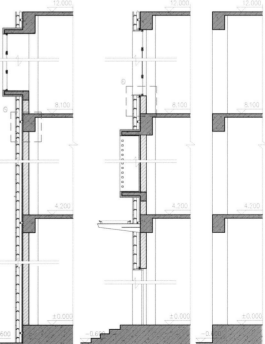

本科二年级

小型公共建筑设计：场地与建筑（6周）

基于城市风景区的自然地形与景观条件，选择若干处坡地完成一个小茶室的设计，主要关注建筑场地与建筑空间的生成、设计方案的深化过程、建筑设计表达方式的综合运用。

Small Public Building Design: Site and Building (6 weeks)

Based on the natural park, a sloping field is chosen to design a tea house on it. This exercise focuses on the relationship of site and building, the design sequence and the comprehensive use of design tools.

环境分析与空间划分
Site Analysis & Space Organization

第一周：制作1:200纸质模型帮助研究分析场地地形条件，主要是景观朝向与场地标高变化，多方案构思空间并以模型表达。

第二周：通过A3图幅1:200比例的平面与剖面草图，在确定方案的基础上调整标高、空间划分、功能流线与场地的关系。

1st week: making 1:200 physical model to support the analysis of topography conditions, including the view point and site gradient, to find out multiple proposals of space generation.

2nd week: using 1:200 plan and section sketches, to adjust relations of site and spatial division and functional streamline.

空间设计的深化：结构与尺度
Structure & Scale

第三周：制作1:100纸质模型和SketchUp模型来帮助进行建筑空间与场地关系的进一步深化调整。

第四周：通过A3图幅1:100比例的平面与剖面草图，根据功能要求在剖面标高变化、尺度和结构方面对空间进行细化。

3rd week: making 1:100 physical model and SketchUp model to support further adjustment on relation of space and the site.

4th week: using 1:100 plan and section sketches, draw details of floor levels' change, dimensions and structures according to functional requirements.

场地与建筑
Site and Building: Tea House in the Natural Park

二层平面 1:100

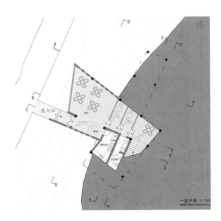

一层平面 1:100

Undergraduate Program 2nd Year

空间质感的表达与构造设计
Material & Construction

　　第五周：使用透视图和模型照片表达空间设计意图，通过模型的透视角度，研究立面材料与构造，绘制1:20墙身剖面图，完成立面与墙身节点的设计。

　　5th week: interpret space design concepts via perspectives, and choose materials and constructions of facade from the perspective point of view. Draw 1:20 wall details to accomplish design of the building.

设计成果的整理与表达
Drawing & Presentation

　　第六周：正式成果图纸、模型的制作，认知排版作为设计的一部分，它与设计意图、设计过程表达的关系。

　　6th week: make formal rendering picture and physical model, and understand the significance of the layout of drawings, as part of design, to design concepts and design process expression.

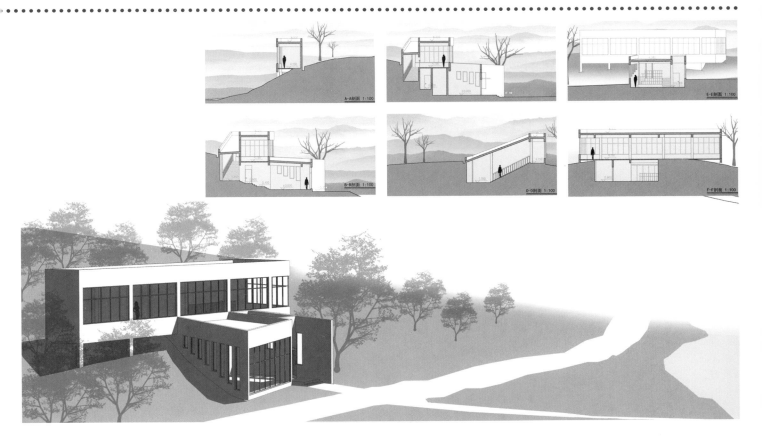

本科三年级

建筑设计（二）小型建筑设计 • 周凌 童滋雨 钟华颖

课程类型：必修
学时学分：72学时/4学分

题目：乡村小住宅客房扩建设计

教学目标

此课程训练最基本的建造问题。通过这一建筑设计课程的训练，使学生在学习设计的初始阶段就知道房子如何造起来，深入认识形成建筑的基本条件：结构、材料、构造原理及其应用方法。同时，课程也面对地形、朝向、功能问题。训练核心是秩序、结构、材料。在学习组织功能与秩序的同时，强化认识建筑结构、建筑构件、建筑围护等实体要素。

任务书

由于发展旅游度假的需要，南京市郊某自然村落村民需要对现有房屋进行加建，改造为小型家庭旅馆。要求在给定基地范围内为小住宅进行扩建设计，为客人提供相对独立的功能齐全的居住与生活设施。目前已经存在的乡村住宅，每户宅基地面积约 90～300 m² 不等，院子面积大小不等，要求在院子内进行扩建设计，建筑面积 180～360 m² 之间，建筑层数 1～3 层，建筑限高 9m，平顶坡顶不限。要求充分考虑材料建造与实施的可能性。

1. 规划布局设计总则
   朝向与视野：主要房间有良好采光、日照，尽量取得良好视野。
   退让：建筑基底与投影不可超出院墙范围内。如若与主体或相邻建筑连接，需满足防火规范。
   边界：建筑与环境之间的界面协调，各户之间界面协调。基地分隔物（围墙或绿化等）不超出用地红线。
   户外空间：每户保持不小于总用地面积10%的户外空间。

2. 建筑单体
   餐厅（10～20人就餐规模）
   厨房区域（≥ 10 m²）
   起居区域（≥ 20 m²）
   学习、娱乐区域（≥ 20m²）
   客房 4～8 间（经济型客房约 15 m²/间，舒适型约 25 m²/间，带卫生间）
   公共卫生间（1～2 间）
   门厅、交通面积酌情设置
   以上各部分功能可开敞布置，也可独立布置。

3. 材料选择
   支撑（承重材料）：木、砖、石、钢、混凝土等
   覆盖（楼地板材料）：木、砖、石、金属、混凝土现浇板等
   围合（墙体材料）：木、砖、石、金属、玻璃等

Subject: Expansion Design of a Rural Small Residential House

Course Objective

This course trains the students to solve the most fundamental problem of construction. Students should learn how to build an architecture at the very beginning of their studying in architectural design, realize the basic aspects which compose architectures: the principles and applications of structure, material and construction. The course also faces the problem of topography, orientation and function. The cores of the course include order, structure and material. Students should strengthen the understanding of physical elements including structures, components and façades while learning to organize the function and order.

Mission Statement

Design the expansion part of the given small house inside the site, providing the individual living accommodation for the guests with complete function. The areas of existing sites are from 90 m² to 300 m² while the existing courtyards have different areas. The expansion should be inside the courtyard with the area from 180 m² to 360 m² and the level from 1 to 3. The limitation of the height is 9 m. The possibility of the material construction and implementation should been fully considered.

1. General Guidelines
Orientation and View: Main rooms should get good sunlight and fine view.
Retreat Distance: The basement & projection of the architecture can't go beyond the courtyard wall. Fire protection rule should be complied.
Boundary: Both the boundary between different buildings and between building and environment should be harmonized.
Open Space: Each site should be more than 10 % of the total area.

2. Built-up Area
Dining Room: for 10～20 people
Kitchen: ≥ 10 m²
Living Area: ≥ 20 m²
Study & Recreation Area: ≥ 20 m²
Guest Room: Economical Room: about 15 m²; Comfortable Room: about 25 m², with toilet
Public Toilet: 1～2 rooms
Lobby and Circulation Space

3. Material
Wood, brick, stone, metal, concrete, glass, etc.

Step 1: 基地与环境 Site and Context

Step 2: 秩序与空间 Order and Space

Step 3: 结构设计 Structure Design

Step 4: 围护设计 Encloser Design

Step 5: 构造设计 Construction Design

Step 6: 细部设计 Detail Design

Step 7: 制图 Drawing

ARCHITECTURAL DESIGN 2: SMALL BUILDING • ZHOU Ling, TONG Ziyu, ZHONG Huaying

Type: Required Course
Study Period and Credits: 72 hours/4 credits

场地 Site

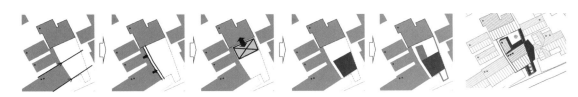

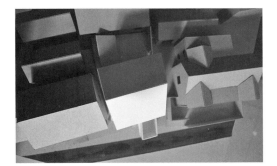

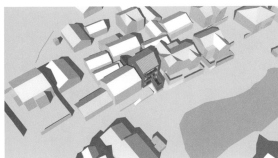

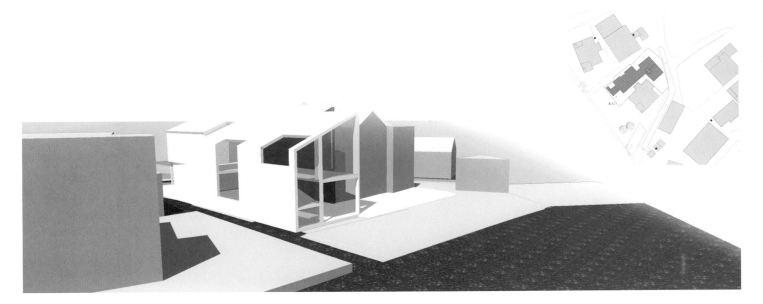

本科三年级

功能与空间 Function and Space

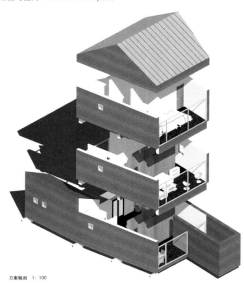

方案轴测 1:100

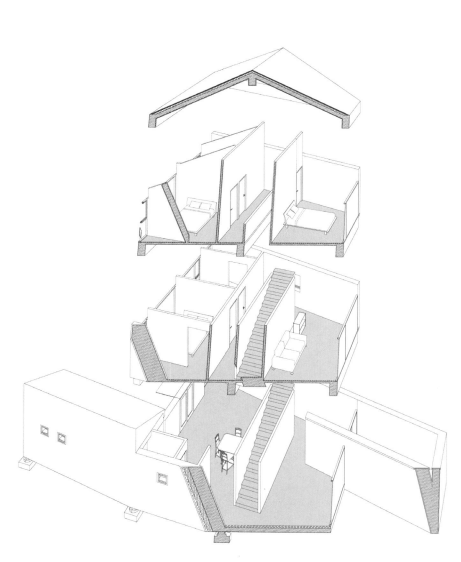

细部 Detail

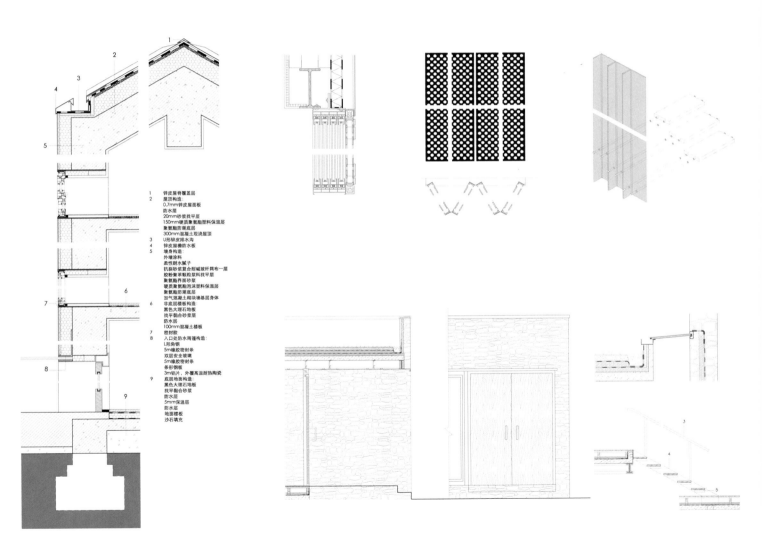

本科三年级

建筑设计（三）中型公共建筑设计 • 周凌 童滋雨 钟华颖

课程类型：必修
学时学分：72学时/4学分

题目：大学生活动中心设计

课程目标

此作业为三年级的第二个作业，设计题目是"大学生活动中心"，作业主题是"空间"。课题训练学生空间组织的技巧，使其掌握空间组织的方法。基地位于校园宿舍区中心轴线花园一角，建筑面积约2500 m²，作业时间8周。要求区分公共与私密空间、服务与被服务空间、开放与封闭空间，训练重点是空间秩序、流线安排、功能配置。要求运用系列剖面、剖透视、轴侧展开图、室内透视图来表达空间。让学生理解图纸不仅是表达工具，也是辅助设计和推敲方案的手段。

课程方法

空间问题是建筑学的基本问题，本课题基于复杂空间组织的训练和学习，从空间秩序入手，安排大空间与小空间，独立空间与重复空间，区分公共与私密空间、服务与被服务空间、开放与封闭空间。训练重点是空间组织，包括空间的秩序、空间的内与外、空间的质感及其构成等。以模型为手段，辅助推敲。设计分为体积、空间、结构、围合阶段，最终形成一个完整的设计。

任务书

1. 空间组织原则

空间组织要有明确特征，有明确意图，概念要清楚，并且满足功能合理、环境协调、流线便捷的要求。注意三种空间：
(1) 聚散空间（门厅、出入口、走廊）；
(2) 序列空间（单元空间）；
(3) 贯通空间（平面和剖面上均需要贯通，内外贯通、左右前后贯通、上下贯通）。

2. 空间类型

(1) 多功能空间：
200座报告厅；容纳80人会议的活动室×2间；容纳40人研讨的活动室×2间
(2) 展示空间：展厅180 m²
(3) 专属空间：
文体类：舞蹈房60 m²×1间；画室60 m²×1间；讲座教学类：教室60 m²×2间
办公类：学生社团活动用房20 m²×8间；教师指导办公用房20 m²×8间
(4) 休闲空间：咖啡座120 m²（附带操作间）
(5) 服务空间：卫生间、储藏间等
(6) 交通空间：门厅、走廊等
总建筑面积控制在2500 m²以内，层数控制在4层以内。

3. 成果

(1) 空间与环境：总平面（1：500）；序列人眼透视（环境融入）。
(2) 空间基本表达：平立剖面（1：200）。
(3) 空间解析与表现：分层轴测；水平楼板秩序轴测；垂直墙体秩序轴测；仰视轴测；剖透视；人眼透视。
(4) 手工模型：1：500总图体量模型；1：300环境模型；1：50单体模型（包含室内空间）

Subject: Design of the College Student Center

Course Objective

This is the second work for Grade - 3 students. The title of the design is College Student Center , and the theme of the work is *space*. This course trains students' skills of space organization and makes them master methods of space organization. The site is located at a corner of the garden on the central axis of the dormitory area in the campus, with a floor space of 2,500 m². The work should be completed in 8 weeks. The public and private spaces, service and serviced spaces, open and close spaces should be differentiated. The key point of the training includes spatial order, circulation arrangement and functional configuration. Students are required to express the spaces using a series of sectional drawings, cutting perspective drawings, shaft-side spreading drawings and indoor perspective drawings. Students should understand that drawings are not only an expression tool, but also a means of assisting design and deliberating plan.

Course Approach

Space issues are the basic issues of architecture. This course organizes trainings and studies based on complex space. Students start with spatial order, arrange the large spaces and small spaces, independent spaces and repetitive spaces, differentiate public and private spaces, service and serviced spaces, open and close spaces. The key point of training is space organization, including the order of space, inner and outer, texture of space and its composition, etc. Models should be used as means to assisting deliberation. The course includes stages of volume, space, structure, enclosing, then forms a complete design.

Mission Statement

1. Principle of space organization

The space organization should have clear characteristics, intentions and concepts, while satisfying the requirements for reasonability, environmental harmony and convenience in circulation. Pay attention to three types of space:
(1) Gathering and dispersing spaces (hallway, entrance / exit, corridor);
(2) Sequence space (unit spaces);
(3) Through space (on plans and sections, including internal/external through, left / right / front / rear through, upper/lower through).

2. Type of space

(1) Multi-function space
Auditorium (200 seats) ×1, activity room (for 80 persons) ×2, activity room (for 40 persons) ×2,
(2) Exhibition space: 180 m²
(3) Special space
Dancing room 60m²×1, drawing room 60m²×1, classroom 60m²×2, office for students association 20m²×8, office for teachers 20m²×8
(4) Recreation space: coffee bar 120 m²
(5) Service space: toilets, storage rooms, etc.
(6) Circulation space: lobby, corridors, etc.
The total floor area ≤ 2,500m², ≤ 4 floors.

Step 1: 基地与环境
Site and Context

Step 2: 空间意象
Space Images

Step 3: 空间操作
Space Operation

Step 4: 环境植入
Environmental Implant

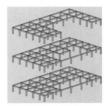

Step 5: 功能置入
Function Implant

Step 6: 结构设计
Construction Design

Step 7: 制图
Drawing

Undergraduate Program 3rd Year
ARCHITECTURAL DESIGN 3: PUBLIC BUILDING • ZHOU Ling, TONG Ziyu, ZHONG Huaying

Type: Required Course
Study Period and Credits: 72 hours/4 credits

概念与原型 Idea and Prototype

"编织"演绎折片　　并排反向交错

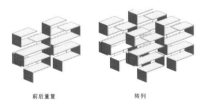

前后重复　　阵列

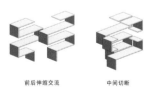

前后伸缩交流　　中间切断

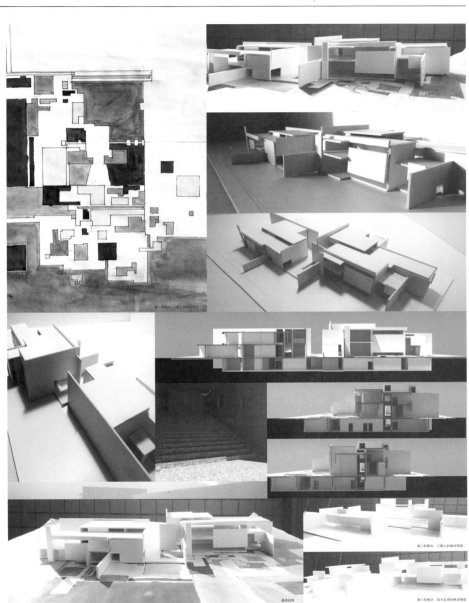

本科生三年级

场地与功能 Site and Function

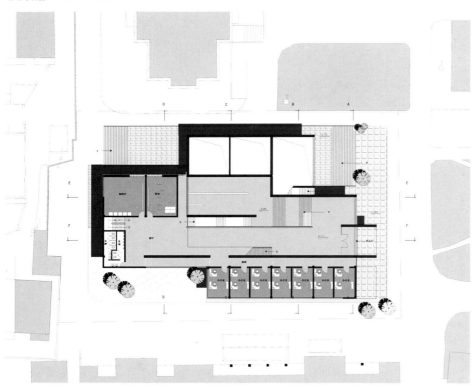

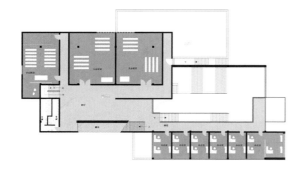

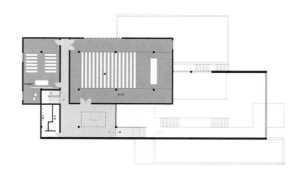

Architecture Program of SAUP, NJU 2011—2012

Undergraduate Program 3rd Year

空间操作 Space Operation

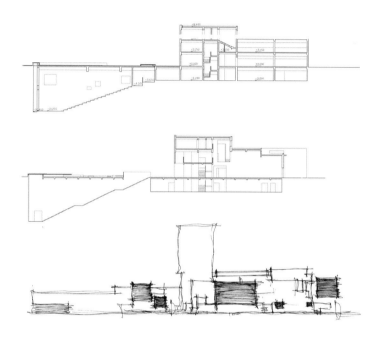
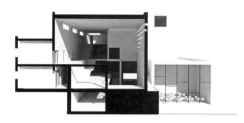
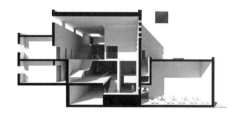
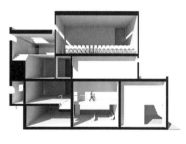
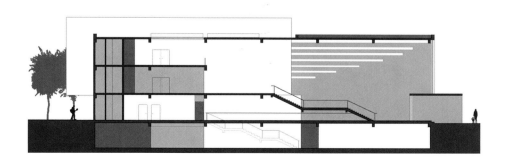
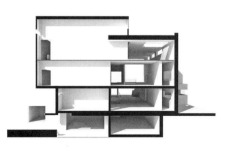

本科三年级

建筑设计（四）大型公共建筑设计 • 华晓宁 胡友培 王丹丹

课程类型：必修
学时学分：72学时/4学分

题目：社区商业中心设计

课程目标

本课程主题为"城市建筑"，希望学生能够在三年级上学期建筑设计课程训练的基础上，进一步将驱动建筑方案生成、发展的外在动力与内在动力进行深入的对接与整合，掌握在复杂的城市场址上组织复杂空间、复杂功能和复杂流线的方法，并为后续的住区规划和城市设计的训练打下基础。

本课程的基本教学思路是从外而内，首先从城市空间入手，将其作为生成建筑的首要驱动力。为此，本年度的任务书改变了传统的建筑设计课程任务书仅仅罗列房间名称和面积的做法，首先要求学生在场地上创造属于城市的外部空间，以外部空间界定建筑形体，并进一步将外部空间系统与内部空间系统深入整合。

任务书

地块位于南京市中央路与南昌路十字路口西南侧，南邻时代超市，北侧为建设新村小区，用地红线内面积11180 m²。

1. 规划限定：
沿中央路退让用地红线15 m，其余边界退让5 m，建筑高度不超过24 m。

2. 设计任务：
创造一个能够吸引更多城市市民和顾客来到的城市公共开放空间。
创造一个服务于社区居民的城市公共开放空间。
创造一条串联这两个公共开放空间的路径，并将这个穿行历程与商业活动整合起来。
约2500 m² ~ 3000 m²的社区服务设施，包括居委会（200 m² ~ 300 m²）、文化站（400 m² ~ 600 m²）、社区卫生所（300 m²）、邮电所（100 m² ~ 150 m²）、修理部（100 m²）、储蓄所（100 m²）等。
约15000 m²的商业、服务与娱乐空间，服务于周边住区及城市空间，其业态选择自定。
约2500m²的辅助用房，包括库房（2000 m²）、卫生间、办公室等
停车空间

Subject: Commercial Center for the Community

Course Objective

The theme of this course is *"Urban Architecture"*. Based on the training of design course in the first semester of Grade-3, students should integrate the internal and external power together which drive the generation and evolution of architectural design, learn the methods of organizing complex space, complex functions and complex circulations on complex urban site, thus constructing the foundations for the following courses of the residential district planning and urban design.

The basic idea of this course is from exterior to interior. The primary driving power for the generation of architecture comes from the urban space. So, the mission statement of this semester changes the old way of just listing out the names and areas of varied rooms. It requires students to generate the open space on the site which will be integrated into the urban space first, then define the volume of architectures from the open space, and integrate the system of external space with the system of internal space together.

Mission Statement

The site is located in the southwest side of the intersection of Nanchang Road and Central Road in Nanjing, north to Times Supermarket and south to the Jianshe Xincun. The land area within the red line is 11,180 m².

1. Planning Limit

The retreat distance from the red line along the Central Road is 15m, while along the other boundaries are 5m. The height of the building should be less than 24m.

2. Design Content

Create an open space attracting more citizens and customers.
Create an open space for the community.
Create a path connecting two open spaces, and integrate the process of walking through with the commercial activities.
Community service facilities: 2,500 m² ~ 3,000 m²
Commercial, service and entertainment space: about 1,5000 m²
Auxiliary space: about 2,500 m²
Parking area

Step 1: 场地
Site

Step 2: 外部空间
Exterior Space

Step 3: 流线
Circulation

Step 4: 功能
Function

Step 5: 技术规范
Building Code

Step 6: 造型
Form

Step 7: 表现
Expression

Step 8: 答辩
Presentation

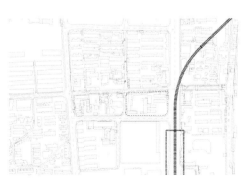

Undergraduate Program 3rd Year

ARCHITECTURAL DESIGN 4: COMPLEX BUILDING • HUA Xiaoning, HU Youpei, WANG Dandan

Type: Required Course
Study Period and Credits: 72 hours/4 credits

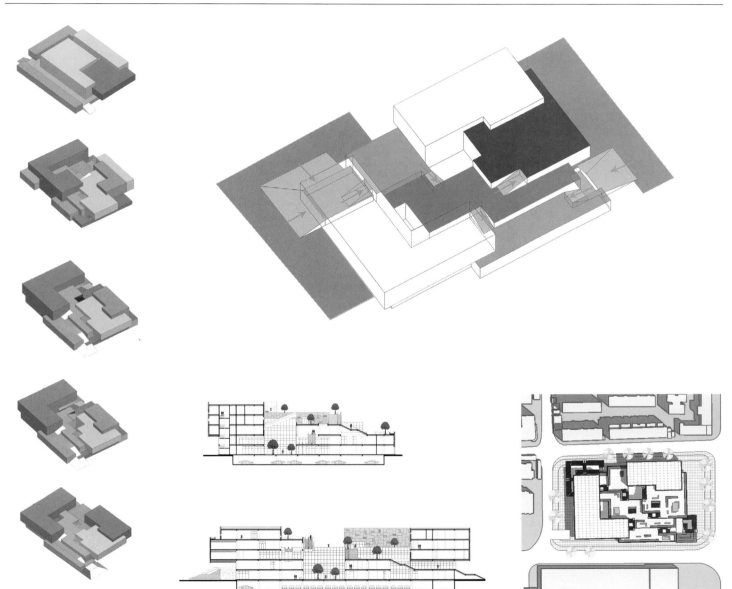

本科三年级

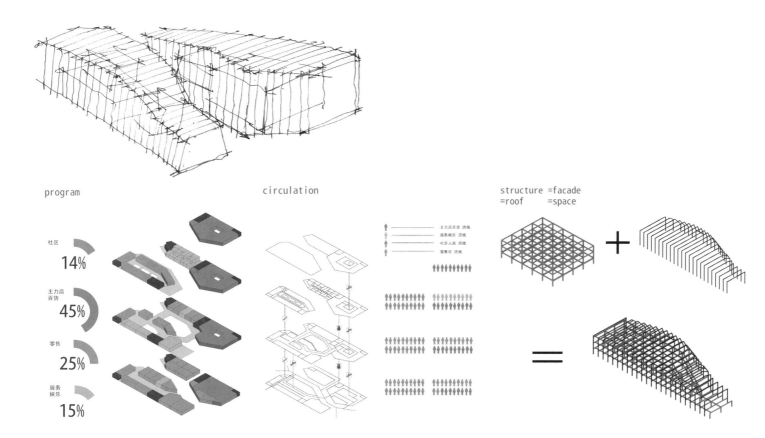

Architecture Program of SAUP, NJU 2011—2012

Undergraduate Program 3rd Year

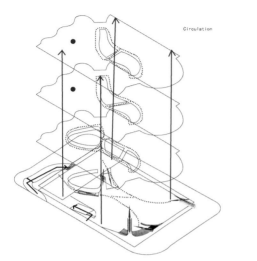

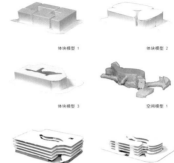

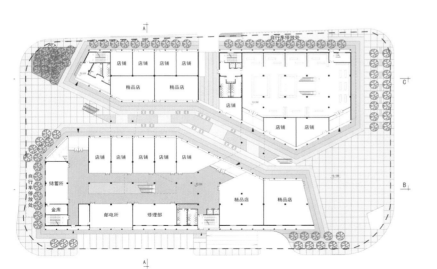

本科三年级

建筑设计（五）住宅小区规划设计 • 华晓宁 胡友培 王丹丹

课程类型：必修
学时学分：72学时/4学分

题目：建设新村地块规划设计

课程目标

本课程设计通过进一步深入研究居住空间与居住行为，要求学生学习建立人性化、体系化的户型系统，并学习掌握住宅区规划设计的基本要求和基本方法。

同时，本课程强调与前一课程题目"社区商业中心"的整合，两者场址相邻，提示学生在完成设计任务的过程中思考和回应两者相互之间以及与周边城市环境的紧密关系，从而凸显两者"城市"这一主题。

此外，本学期本课程的教学强调了与建筑物理、建筑设备课程的整合，以城市住宅为载体，使学生初步掌握与建筑节能和可持续设计相关的知识与技能。

任务书

地块位于南昌路以北、中央路以西、芦席营路以东，面积约65000 m²。地块与建筑设计（四）任务地块相邻。学生需充分考虑与周边环境以及南侧社区商业服务中心的联系。

1. 主要规划指标
   用地性质：R2　　用地面积：65000 m²
   容积率：1.7～1.8　建筑高度：H ≤ 35 m
   建筑密度：≤ 30％　绿地率：≥ 35％
   商业面积：≤5000 m²
2. 退让
   高层退让中央路红线25 m，退其他道路15 m；低、多层退让中央路15 m，退其他道路5 m。
3. 户型
   类型：小高层
   面积主要分为4档：90 m²左右，120 m²左右，140 m²左右，140 m²以上
   面积配比：90 m²以下户型比例要求不少于30％，140 m²以上户型不超过30％。
4. 配套公建
   主要包括：4班幼儿园、会所（1000 m²）、沿街商业（沿中央路布置，不超过两层）。
5. 机动车位配置
   户型建筑面积在140 m²以上的，1车位/户；90～140 m²的，0.7车位/户；<90 m²的，0.5车位/户。地上机动车停车20％，需对地下空间利用进行规划。

Subject: Design of the Jianshe Xincun Block

Course Objective
In this course, students should study the living behaviors and activities, learn to establish humanized system of residential units, understand the basic requirements and methodology of the planning of residential district.
Meanwhile, this course emphasizes the integration with the former course "Commercial Center for the Community". Two adjoining sites suggest students consider and respond the close relationship between two projects and the surrounding urban space, thus emphasizing the urban theme.
Furthermore, in this semester the integration of this course with the courses of Building Physics and Building Equipment have been emphasized. With the urban residence as the carrier, students should learn the basic knowledge and techniques of building energy efficiency and sustainable design.

Mission Statement
The site is located in the north of Nanchang Road, west of Central Road, east of Luxiying Road, with an area about 65,000 m². It is adjacent to the site of Architectural Design Course 4, which suggest students to fully consider the relationship with surrounding environment and the southern commercial center for community.
1. Main planning indicators
The nature of land: R2　　Land area: 65,000 m²
Plot ratio: 1.7 ~ 1.8　　Building height: H ≤ 35 m
Building density: ≤ 30 %　Green space ratio: ≥ 35 %
Commercial area: ≤ 5,000 m²
2. Retreat Distance
High buildings retreat 25m from red line of Central Road, 15m from other roads; low and multi-story buildings retreat 15m from the Central Road, 5m from other roads.
3. Unit
Building type: high-rise buildings
Unit area types: about 90 m², about 120 m², about 140 m², more than 140 m²
Area ratio: ≤ 90 m² : more than 30%
　　　　　≥140 m²: less than 30%.
4. Public buildings
Kindergarten of 4 classes, clubs (1,000 m²), commercial building (along the Central Road, 1~ 2 layers).
5. Vehicle parking configuration
Those which building area is more than 140 m², a parking / household; those 90 ~ 140 m², 0.7 parking / household; those < 90 m², 0.5 parking / household. Parking of motor vehicle on the ground is 20%, it is necessary to plan the use of underground space.

Step 0: 调研 Investigation

Step 1: 户型 Unit

Step 2: 单体 Building

Step 3: 规划结构 Planning Structure

Step 4: 交通系统 Traffic System

Step5: 物理环境 Physical Environment

Step 6: 景观 Landscape

Step 7: 公建 Service Building

Step 8: 表现 Expression

Step 9: 答辩 Presentation

… Undergraduate Program 3rd Year
ARCHITECTURAL DESIGN 5: RESIDENTIAL PLANNING • HUA Xiaoning, HU Youpei, WANG Dandan

Type: Required Course
Study Period and Credits: 72 hours/4 credits

规划结构 Planning Structure

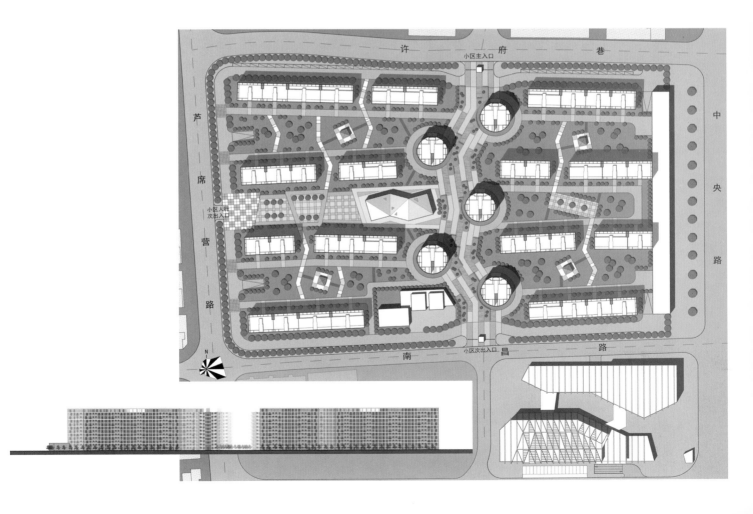

本科三年级

城市关系 Urban Relationship

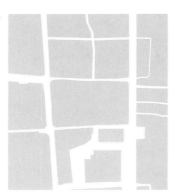 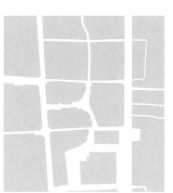 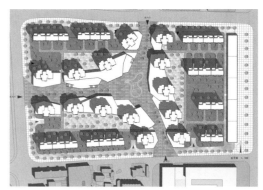

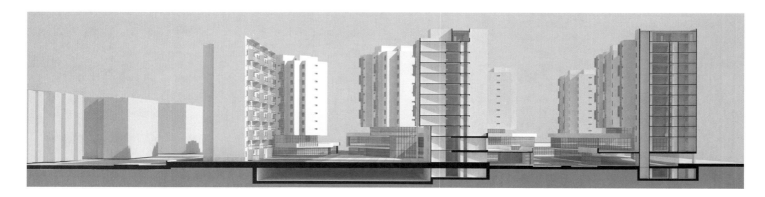

Architecture Program of SAUP, NJU 2011—2012

Undergraduate Program 3rd Year

物理环境　Physical Enviroment

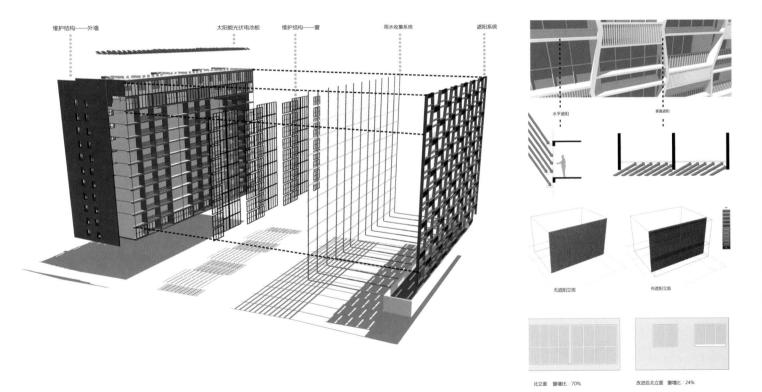

本科四年级

建筑设计（六）城市设计 • 丁沃沃 尹航 胡友培

课程类型：必修
学时学分：72学时/4学分

课程介绍
在城市化进程不断加剧的社会发展阶段，城市建筑的性质也发生了变化，建筑不再是以自我为中心的独立个体，而更加融入周边的整体环境。因此，认知建筑新的角色、理解建筑在城市环境中的位置，是当今建筑师必备的知识。本次练习主要关注城市更新和城市再生问题，同时关注城市中自然元素的修复与利用。通过城市设计练习认识城市建筑的本质，以及建筑与城市的关系。

课程目标
1. 熟练掌握城市设计的方法，熟悉从宏观整体层面处理不同尺度空间的能力，并有效地进行图纸表达。
2. 通过调查，理解城市更新的概念和价值；通过分析理解城市交通、城市设施在城市体系中的作用；通过研究理解城市形态与城市功能之间的内在关系，认识城市肌理与城市规划指标体系之间的内在关系；通过设计掌握城市建筑的基本规律，实践处理群体建筑问题的要义；通过体验认知城市公共开放空间与城市日常生活场所的关联，运用景观环境的策略创造城市空间的特征。
3. 多人小组合作，培养团队合作意识和分工协作的工作方式。

Course Introduction
This urban design studio would help students make deep understandings about the essence of urban architecture. In a social stage with rapid urbanization, the role of architecture has been transformed, shifting from individual objects to a part of an integrated urban environment. So, the knowledge about the new role of architecture and its position in urban environment becomes critical and necessary. The domain of urban design is diversified. However, this studio focuses on urban renewal and re-intervention with further emphasis on the restoration and utilization of nature elements in urban areas.

Course Objective
1. Training students to master urban design methods, to design in diversified urban scale with an overall urban vision of the project, to present the design in a professional way.
2. Training students to understand the idea and the value of urban renewal through field survey; understand the role of urban traffic and infrastructure in urban system through analysis; to understand the internal relationship between urban fabric and plot indicators through study on the link between urban form and function; Training students to be familiar with the basic rules of urban architecture through design practice, to exercise on the ability of designing group buildings. Training students to construct a knowledge link between urban open space and urban life, further develop landscape strategies to create special urban places based on their daily life experiences.
3. Team works, training the ability of cooperation and negotiation.

第一周 Week 1	第二周 Week 2	第三周 Week 3	第四周 Week 4	第五周 Week 5	第六周 Week 6	第七周 Week 7	第八周 Week 8
基地介绍与场地调研 Survey and Mapping		城市公共空间与肌理 大街区总图规划设计 Urban Fabric and Master Plan		城市肌理与建筑类型 Urban Fabric and Building Type		表现图练习 Representation	
	城市形态结构性框架 Structural Planning		城市交通与场地交通组织 大街区总图优化 Traffic, Circulation and Master Plan		场地环境与公共空间深化设计 Open Space and Landscape		答辩 Presentation

Undergraduate Program 4th Year

ARCHITECTURAL DESIGN 6: URBAN DESIGN • DING Wowo, YIN Hang, Hu Youpei

Type: Required Course
Study Period and Credits: 72 hours/4 credits

形态结构框架设计
Structural Planning

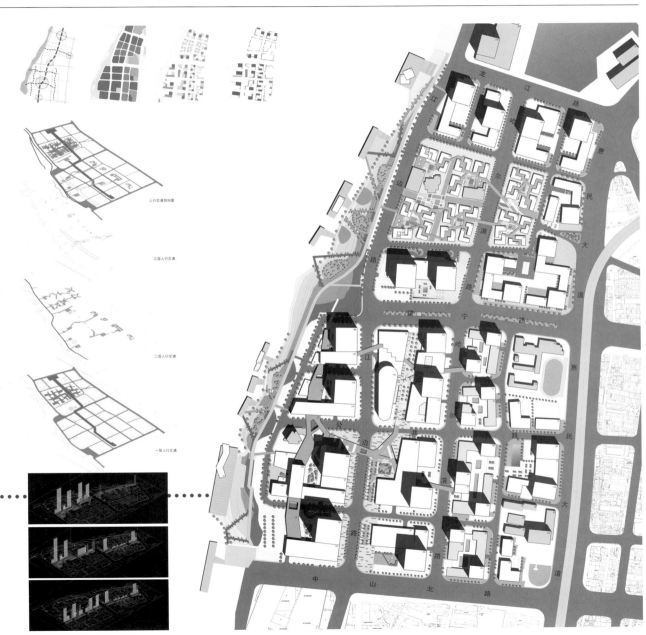

建筑类型与开放空间
BuildingType and Open Space

Undergraduate Program 4th Year

景观设计
Landscape Design

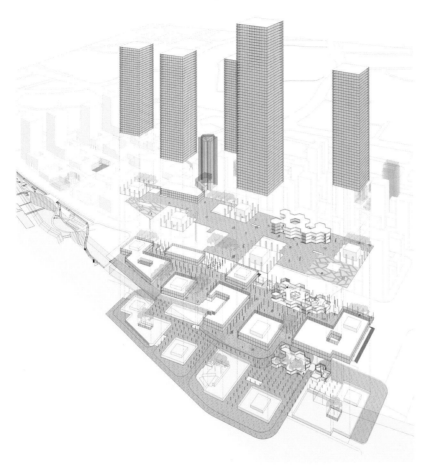
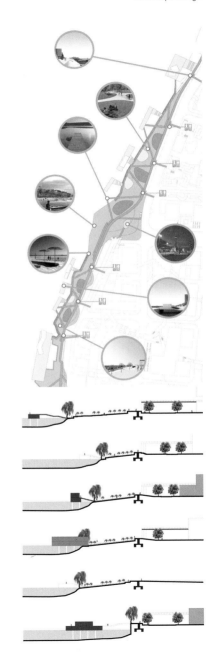

本科四年级

建筑设计（七） 高层建筑设计 • 吉国华 胡友培 尹航

课程类型：必修
学时学分：72学时/4学分

课程目标
高层办公建筑设计涉及城市、空间、形体、结构、设备、材料、消防等方面内容，是一项较复杂与综合的任务。本课题采取贴近真实实践的视角，教学重点与目标是帮助学生理解、消化以及涉及各方面知识，掌握综合运用知识并富有创造性的解决问题的技能。

项目概况
本选址位于白下路内桥西南，东临中华路，北、西、南三面分别为原基督教青年会旧址（文物保护单位）、南京第一中学以及江苏国际经贸大厦。

规划要点及控制条件
根据规划设计要点，本项目建筑容积率≤7，建筑高度≤160m，裙房高度≤24m；文物建筑保护范围内不得进行新的建设，南侧控制地带内允许建设多层建筑。

规划高层建筑与周边被遮挡现状住宅和学校的建筑间距应满足日照计算的要求，同时规划建筑与周边现状建筑的间距以及退让道路红线和用地边界应符合《南京市城市规划条例实施细则（2007版）》的有关规定要求。

功能要求
高层部分为办公楼，设计应兼顾各种办公空间形式。

裙房主要功能为餐馆，应合理组织餐厅与厨房关系，餐厅部分须设置800 m²大型宴会厅1个和400 m²中型宴会厅1个

地下部分主要为车库和设备用房。汽车库和自行车库的配置应满足《南京市建筑物配建停车设施设置标准与准则》的要求。

Course Objective
The design of high-rise office building is a complex and integrative work which concerns city, space, volume, structure, equipment, material, fire protection, etc. This course takes the viewpoint of the real practice. The objective and key point of the course is to help the students understand and digest the above knowledge, thus learn the skills of solving problems creatively using the knowledge comprehensively.

General Introduction
The site is located in the southwest side of the Neiqiao on Baixia Road. The north side of the site is the former site of the YMCA (Young Men's Christian Association). The west side of the site is the No.1 Middle School. The south side of the site is the Jiangsu International Trade Building.

Planning Indicator
Plot ratio: ≤7
Height of the building: ≤ 160 m   Height of the skirt building: ≤ 24 m.
New construction is forbidden in the area of historical architectures' protecting.
The retreat distance should satisfy the requirement of sunlight and other rules.

Functional requirements
The high rise should be the office room considering the different types of office space. The skirt building should be the restaurant with good relationship between the dining rooms and the kitchens, including a big banqueting hall of 800 m² and a middle banqueting hall of 400 m².
The underground space should be the parking area and the equipment rooms. The configuration of the parking area should be according to the national and local rules.

Undergraduate Program 4st Year

ARCHITECTURAL DESIGN 7: HIGH-RISE BUILDING • JI Guohua, HU Youpei, YIN Hang

Type: Required Course
Study Period and Credits: 72 hours/4 credits

场地与形态 Site and Volume

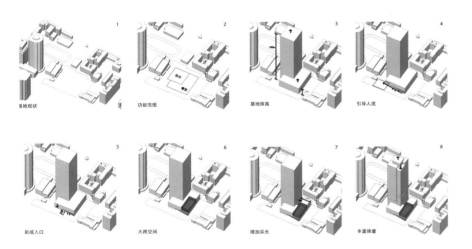

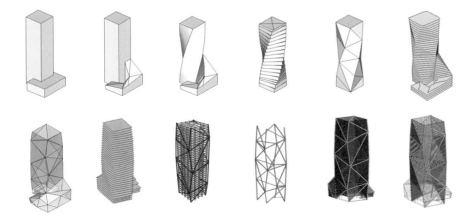

本科四年级

功能与空间 Function and Space

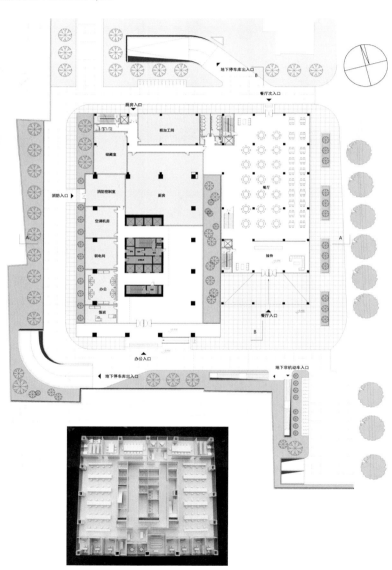
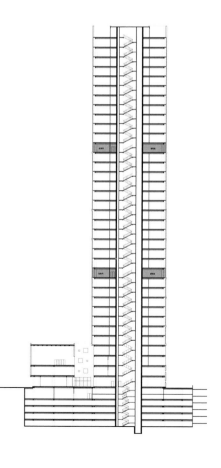

表皮与细部 Facade and Detail

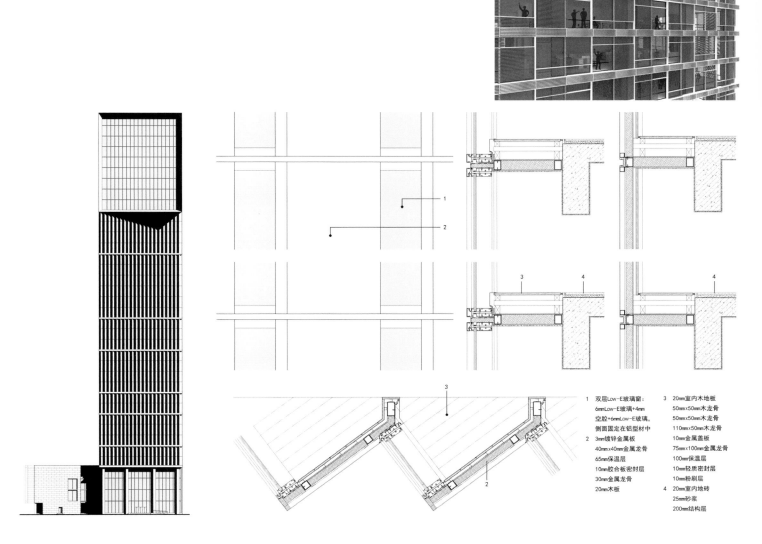

1. 双层Low-E玻璃窗：
   6mmLow-E玻璃+4mm
   空腔+6mmLow-E玻璃，
   侧面固定在铝型材中
2. 3mm镀锌金属板
   40mm×40mm金属龙骨
   65mm保温层
   10mm胶合板密封层
   30mm金属龙骨
   20mm木板
3. 20mm室内木地板
   50mm×50mm木龙骨
   50mm×50mm木龙骨
   110mm×50mm木龙骨
   10mm金属盖板
   75mm×100mm金属龙骨
   100mm保温层
   10mm轻质密封层
   10mm粉刷层
4. 20mm室内地砖
   25mm砂浆
   200mm结构层

研究生一年级
建筑设计研究（一）基本设计 • 张雷

课程类型：必修
学时学分：54学时/2学分

题目：南京城南大板巷西侧、绫庄巷两侧更新改造设计研究

课程目标
课程从"环境"、"空间"、"场所"与"建造"等基本的建筑问题出发，通过南京老城南城市肌理和建筑类型的分析以及其后功能置换后使用空间的重新划分，从建筑与基地、空间与活动、材料与实施等关系入手，将问题的分析理解与专业的表达相结合，达到对建筑设计过程与设计方法的基本认识与理解。

研究主题
建筑类型 / 空间再划分 / 建筑更新 / 建造逻辑

设计内容
对老城南大板巷西侧、绫庄巷两侧进行调研，每组选择一个院落或区域，通过功能置换和整修改造，使其满足新的使用要求。

Subject: Renovation of the West Side of Daban Lane and Two Sides of Lingzhuang Lane

Course Objective
Based on the basic architectural problems such as environment, space and place, and construction, the course asks students to begin with analyzing the relationship of buildings and the site, space and behavior, materials and implementation, to understand the old city fabric and later displaced function, and combine the professional expression with the analysis of the architectural problems, so as to comprehend the basic architectural design process and design method.

Research Subject
Architectural Types / Space Redefine / Architectural Renovation / Tectonic Logic

Design Content
Do investigation on the west side of Daban Lane and both sides of Lingzhuang Lane. Each group chooses one court or region, make it satisfy the new requirement through the functional replacement and reconstruction.

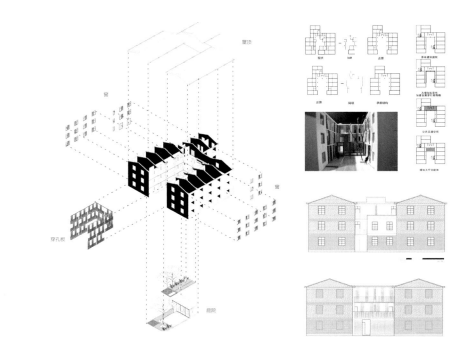

Architecture Program of SAUP, NJU 2011—2012

Graduate Program 1st Year
DESIGN STUDIO 1: BASIC DESIGN • ZHANG Lei

Type: Required Course
Study Period and Credits: 54 hours/2 credits

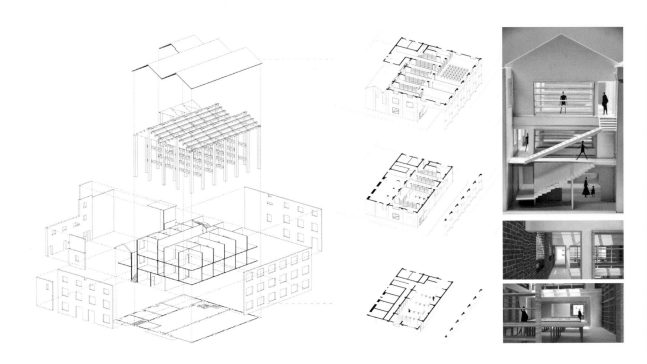

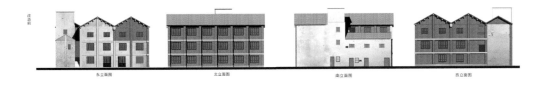

研究生一年级
建筑设计研究（一） 基本设计 • 傅筱

课程类型：必修
学时学分：54学时/2学分

题目：某设计院建筑创作空间扩建

课程目标

课程从"空间"、"场所"与"建造"等建筑的基本问题出发，通过某设计院创作空间的扩建，着重训练学生对建筑与基地、空间与行为等关系的认知，从而加深对建筑设计过程与设计方法的基本认识。

研究主题

建筑形态与周边环境/空间构成与行为模式

设计内容

在原有设计院北面扩建约3300 m²的建筑创作空间（包含建筑师个人工作室、集体创作室、模型制作车间等）。

Subject: Expansion of Design Studio for An Architectural Design Institute

Course Objective

The course starts from some basic items, such as space, site and construction, and focuses on enhancing students' awareness of relations between building and construction sites, spaces and behaviors via expansion of design studio for an architectural design institute, in order to help students get a basic understanding on building design process and methods.

Research Subject

Structure of building form to surrounding environment / spaces and behavior model

Design Content

Expansion of 3,300m² space on the north of existing design institute, which may include individual studios for the architect, team work space and model manufacturing room.

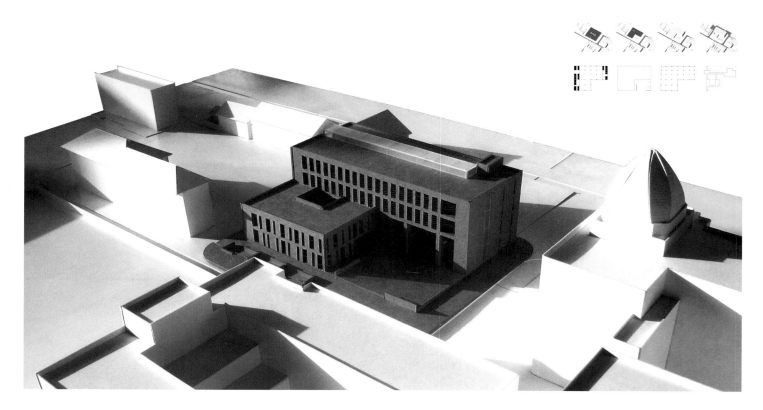

Graduate Program 1st Year
DESIGN STUDIO 1: BASIC DESIGN • FU Xiao

Type: Required Course
Study Period and Credits: 54 hours/2 credits

Teaching Process

1. Site analysis
Time: 1st week
Content: analyze the site and submit the analysis report and site model (1 per group, 1:200).
2. Understand basic architectural design principles
Time: 2nd week
Content: give a lesson with case study, and carry out discussion.
3. Organize space and behavior
Time: 3rd, 4th, 5th and 6th weeks
Content: conceive, discuss and form solutions, and study and discuss them based on a single working model and PPT. The form of expression on the drawings at this stage emphasizes conceptual expression.
4. Design study and expression
Time: 7th and 8th weeks
Content: complete the design documents, a 1:20 plan detail of wall and a 1:200 single working model. The form of expression on the drawings at this stage emphasizes engineering expression.

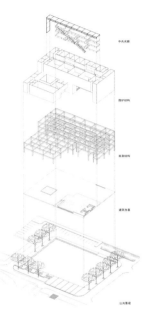

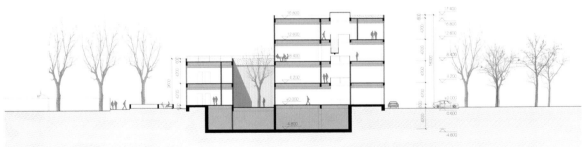

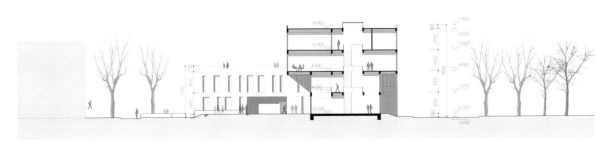

研究生一年级
建筑设计研究（一） 概念设计 • 周凌

课程类型：必修
学时学分：54学时/2学分

题目：垂直城市

课程目标

此课题致力于探索中国未来都市的形象。中国城市发展，实践先于理论是一个长期存在的事实。一方面城市化加剧，城市快速扩张；另一方面城市无序和混乱状况继续存在。我们需要探索中国未来城市建筑的可能性。新的城市建筑应该具有更加合理的功能，便捷的交通，同时具有归属感，具有精神意义，包含某种新的纪念性。

研究主题

提出一个垂直城市的概念。垂直城市必须是高效的、混合的、生态的，而且具有一定地面感。

设计内容

垂直城市可以是垂直社区、商场、校园等，包含居住、办公、集会、文化、休憩、娱乐以及餐饮、购物、休闲等生活设施的综合体。地点选择在南京城内，垂直城市不仅要安置地块原有功能，还要增加至少一倍的建筑面积，让出一倍土地面积作为城市绿地。

教学过程

第1周 阅读理论；
第2周 阅读事件；
第3周 阅读城市；
第4~8周 概念设计：三人一组，分别以科学家、艺术家的方式工作，首先分别理性和创造性思维进行提案设计，然后交流合作，共同完成设计。

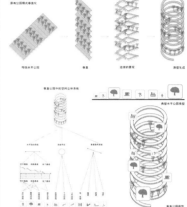

Subject: Vertical City

Course Objective

The course is to explore what China's cities will look like in the future. It is a timeless fact that practice goes ahead of theory. On one hand, urbanization happens faster and the range of the city is getting wider. On the other hand, disorder and chaos still exist in cities. Therefore, we need to find out possibilities for the future of urban construction in China. Buildings in new cities shall have more reasonable functions, convenient traffic conditions, where citizens feel more sense of belonging and spiritual meaning, even including some new monumental significance.

Research Subject

Introduce the concept of one vertical city, that shall be highly efficient, mixed, ecological, and with a sense of ground surface.

Design Content

The vertical city can be a sort of vertical community; emporiums and campuses, an integrated place covering the functions of living, working, meeting, culture, rest, entertainment, catering, shopping and leisure. The place is in downtown of Nanjing, where the vertical city shall not only maintain the existing functions on the land parcel, but also have building areas expanded as much as twice of the previous building areas, and the saved land shall be used as green space.

Teaching Process

1st week: read the theory;
2nd week: read the events;
3rd week: read the city;
4th~8th weeks: conceptual design by groups consists of three members for each group. The members of each group shall work as scientist or artist respectively, to make draft design from the sense and creation thinking point of view. Afterwards, the members of each group shall communicate and cooperate with each other to accomplish the design jointly.

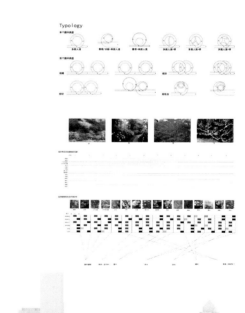

垂直公园
Vertical Park

DESIGN STUDIO 1: CONCEPTUAL DESIGN • ZHOU Ling

Type: Required Course
Study Period and Credits: 54 hours/2 credits

垂直校园
Vertical Campus

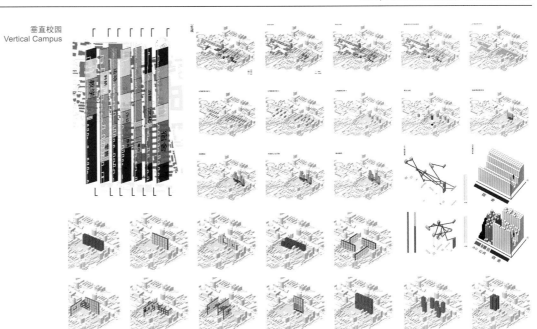

研究生一年级
建筑设计研究（一） 概念设计 • 张旭 刘可南

课程类型：必修
学时学分：54学时/2学分

题目：停车塔楼更新

课程目标
作为一个故事发生的场所，停车场往往出现在小说家、摄影家和电影导演的作品中。而作为一个构筑物，停车场却处于城市文化的边缘。它用巨大的尺度沉默地定义了城市空间，却也经常性地被人遗忘。本次设计以停车场作为一个契机，来重新审视城市的现状和未来的可能性。

研究主题
　　城市环境/功能混用
　　结构体系/空间界面/细部设计

设计内容
　　为位于香港的一处停车楼寻找新的使用可能性。并且在原有建筑物基础上进行改造，使其重获新生。

Subject: Renew of Parking Tower

Course Objective
As a place full of stories, the parking area is always mentioned in articles written by novelists, photos taken by photographers and movies edited by directors. In addition, as a structure, the parking area is in the edge of city culture. Parking area silently defined the city spaces in a huge dimension, but is easily forgotten by people. The design takes the parking area as the breakthrough point to re-scan the current conditions and future possibilities of city.

Research Subject
City environment/function mixing
Structural system/space interface/detailed design

Design Content
Find out new possibilities for one parking tower in Hong Kong to give the old building a new life.

Graduate Program 1st Year

DESIGN STUDIO 1: CONCEPTAL DESIGN • ZHANG Xu, LIU Ke'nan

Type: Required Course
Study Period and Credits: 54 hours/2 credits

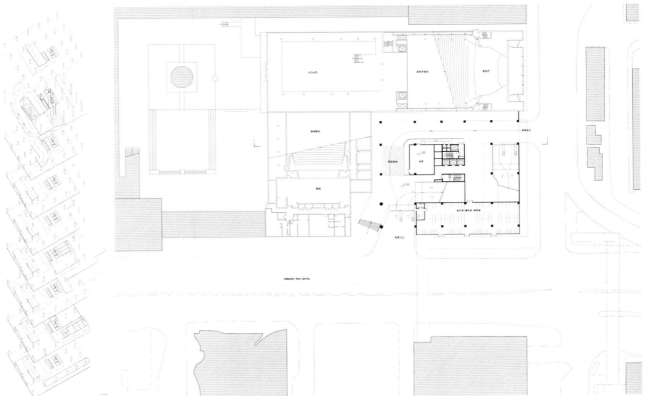

研究生一年级
建筑设计研究（二） 建构设计·傅筱

课程类型：必修
学时学分：54学时/2学分

题目：《基础设计》的深化与发展

课程目标
  1. 不同结构类型对应的空间形态特征研究
  2. 设计概念与构造设计

设计内容与计划（2人/组）
  1. 结构类型转换
    根据基础设计的结构类型转换为另外一种不同的结构类型，重点研究不同结构类型产生的空间特征，并研究结构类型对原设计概念的制约或促进关系。
    时间：3周
    成果：实物比例模型，概念分析图，设计图纸
  2. 设计概念与构造设计
  （1）处理好节点的基本工程技术问题
    a. 对自然力的抵抗与利用：保温、防水、遮阳……
    b. 构造的施工：复杂问题简单化，建造方便性，误差
  （2）根据设计概念研究建造材料的选用和节点设计，在满足基本功能性构造技术的前提下，重点研究超越功能性技术问题的构造设计表达。
    时间：4周
    成果：数字模型、概念分析图、构造详图
  3. 设计成果整理与表现
    大比例平立剖图纸表达深度达到施工图深度
    节点大样必须有剖面节点、平面节点、三维节点示意，必须有带节点的空间透视表达，必须有带节点的三维轴测表达
    时间：1周
    成果：PPT演示和A1展板（不少于2张）

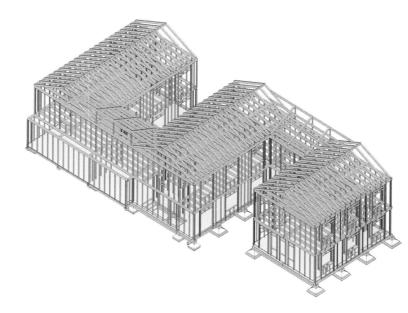

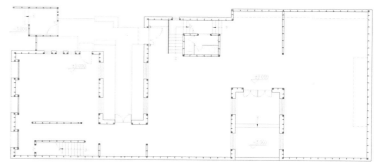

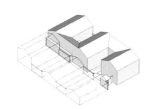

Graduate Program 1st Year

DESIGN STUDIO 2: CONSTRUCTIONAL DESIGN • FU Xiao

Type: Required Course
Study Period and Credits: 54 hours/2 credits

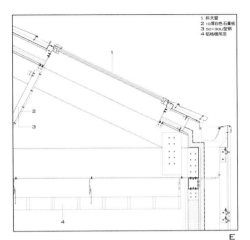

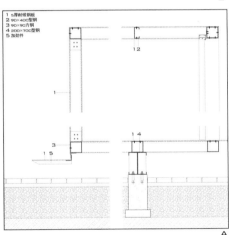
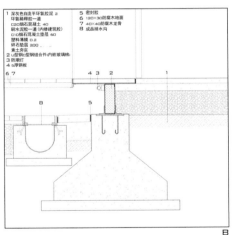
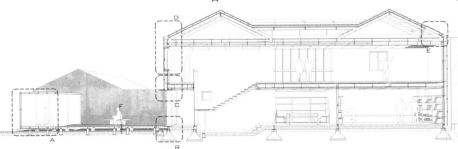

Subject: Design Development of Studio 1 " Basic Design "

Course Objective
1. Study on the morphological characteristics of the space corresponding to different structural types
2. Design concepts and structural design

Design Content and Program
1. Structural type conversion
Requirements: Convert the structural type introduced in "Basic Design" into a different type, focusing on studying the spatial characteristics of space generated by different structural types, as well as studying the restriction or promotion of structural type to the original design concept.
Duration: 3 weeks
Achievements: physical scale models, conceptual analysis diagrams, design drawings
2. Design concept and structural design
Requirements:
(1) Settle the basic engineering technical issues regarding the nodes
a. Resistance to and utilization of the natural force: heat preservation, waterproofing, sun shading, and more.
b. Construction of structure: simplification of complex problems, construction convenience, error issues…
(2) Based upon the design concepts, study the selection of construction materials and the design of nodes; and on the premise of satisfying basic functional construction techniques, place emphasis on studying the expression of structural design beyond technical issues.
Duration: 4 weeks
Achievements: digital models, conceptual analysis diagrams, detailed constructional drawings
3. Reorganization and expression of design results
Requirements: The depth of expressions on large-scale plan drawings, elevation drawings and sectional drawings shall reach that of the construction drawings; the detailed drawing of nodes must have sectional nodes, plan nodes and 3-dimensional nodes; the expression must have spatial perspective expressions with nodes and 3-dimensional isometric expressions with nodes.
Duration: 1 week
Achievements: PPT presentations and A1-format exhibition boards (at least 2 pieces.)

研究生一年级

建筑设计研究（二） 建构设计 • 郭屹民

课程类型：必修
学时学分：54学时/2学分

课程目标
1. 掌握结构设计基础知识，并会进行结构分析和结构设计
2. 了解结构设计与功能的关系，并会进行与功能相关的建筑结构设计
3. 了解结构的材料与建造，并会通过材料和建造进行建筑结构设计

课程内容
1. 结构基础
2. 结构发展史
3. 结构设计与形态建构
4. 结构方法与建筑设计

课程作业
1. 结构设计与形态建构案例分析（两人一组）
2. 大跨度结构设计（两人一组）
3. 南大建筑毕业设计校园展廊（两人一组）

Course Objective
1. Learn the basic knowledge of structural design. Learn to analyze and design the structure of architecture.
2. Understand the relationship of structural design and function. Learn to design the structure of architecture according to the function.
3. Understand the material and construction of structure. Learn to design the structure of architecture according to the material and construction.

Course Content
1. Basic of structural design
2. Types and history of structure
3. Relationship between structural design and form
4. Structural design and architectural design

Exercise
1. Case Analysis on the structural design.
2. Design of the long-span structure (two students per group).
3. Design of a temporary gallery for the graduation design of SA in NJU (two students per group).

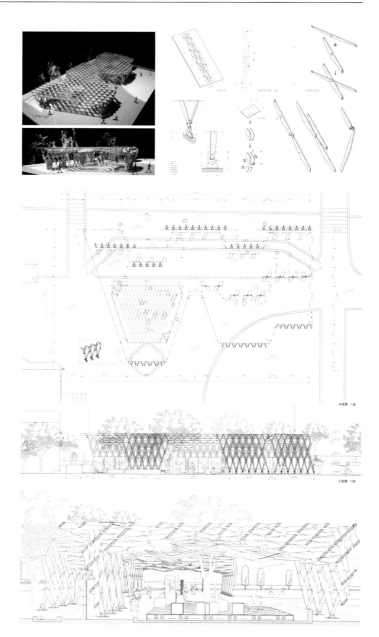

Graduate Program 1st Year
DESIGN STUDIO 2: CONSTRUCTIONAL DESIGN • GUO Yimin

Type: Required Course
Study Period and Credits: 54 hours/2 credits

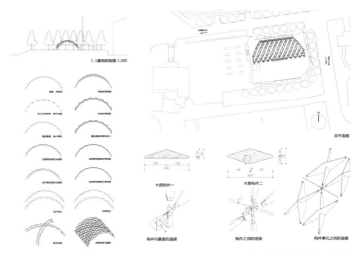

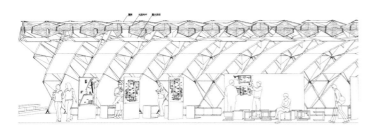

研究生一年级
建筑设计研究（二） 城市设计 • 丁沃沃

课程类型：必修
学时学分：54学时/2学分

题目：多维城市意向

课程目标
　　城市化的结果导致了城市成为人们生存的主要空间，因此，城市的物质空间不仅影响到人们对环境的认知，更势影响到人们对环境恰当的反馈。对城市物质空间的研究当下已经成为建筑学的重要内容。本课题拟通过城市设计的操作实验了解城市物质空间的本质和意义，初步掌握物质空间、指标体系、互联网络、物理环境等要素之间的关系。

研究主题
　　多维城市空间操作的可能性及其方法

设计内容
　　1. 以南京新街口中心区作为设计实验的场所，以城市更新与发展为城市设计的主要主体和动力。
　　2. 观察与发现城市物质空间内的运作现象，图示城市物质空间的运行模式及其问题，通过设计实验提出解决问题的途径，通过设计操作实现方案的设想。

Subject: Multi-dimensional City

Course Objective
Urbanization leads to the result that the city becomes the main living space for people. Therefore, the physical spaces in cities affect not only the people's awareness of environment, but also their suitable feedback to the environment. Research on urban physical spaces is now becoming one of the most important issues in architecture. The course aims to understand nature and the significance of urban physical spaces via the urban design process, and to preliminarily find out the relationship among physical space, index system, Internet and physical environment, etc.

Research Subject
Possibilities and methods to operate multi-dimensional urban space

Design Content
1. It takes the Nanjing Xinjiekou commercial center as the place of trial design, and regards the city update and development as the main subject and driven force of the urban design.
2. It is planned to observe and discover the operation of urban physical space, draw out operation mode and issues of urban physical space, find out ways to resolve the issues via trial design and finally realize the design proposal by design implementation.

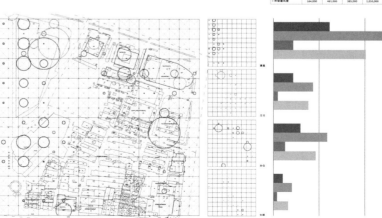

Graduate Program 1st Year
DESIGN STUDIO 2: URBAN DESIGN • DING Wowo

Type: Required Course
Study Period and Credits: 54 hours/2 credits

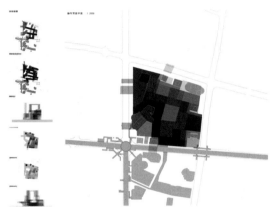
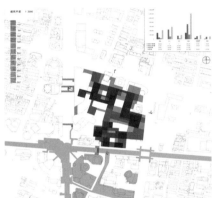
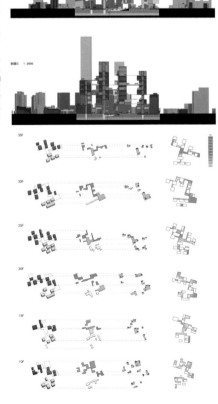
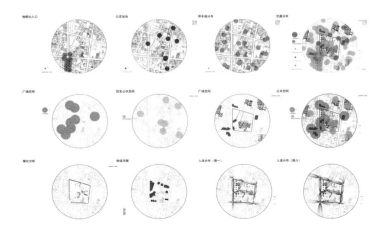

研究生一年级
建筑设计研究（二） 城市设计 • 冯路

课程类型：必修
学时学分：54学时/2学分

题目：上海龙华寺周边地区改造城市设计

课程目标
　　上海龙华寺周边地区改造涉及历史区域保护、城市更新、商业开发、立体交通组织等综合复杂内容，而这正是中国大城市经常面临的情况。通过这个城市设计课程，使学生理解城市设计的意义和设计方法，理解当代城市的复杂状况，学习在复杂状况下处理城市各因素之间的关系，学习制定城市设计导则，并理解城市设计所内含的社会、政治、文化意义。

研究主题
　　建筑与城市、保护与更新、公共空间的塑造

设计内容
　　改造共分5个地块，其中包括保护区域、商业开发和公共绿地等。在现有规划的基础上，对规划提出优化建议；分析改造对周边城市区域的影响；设置和组织各地块建筑功能；建筑设计控制导则；历史保护区域的重点设计；立体交通组织设计；景观空间的塑造。

教学过程
　　共8周，设计过程并非被简单限定为不同步骤，而在于不同阶段之间的叠合。
1. 场地分析
时间：第1周到第3周
内容：分析场地，并提交分析图表和场地模型
2. 案例讨论
时间：第2周
内容：学生分组寻找类似案例，集体学习和讨论
3. 设计概念的产生
时间：第3周到第5周
内容：构思、讨论、形成设计概念，以图纸和工作模型进行研究讨论
4. 设计概念的落实和完善
时间：第5周到第7周
内容：通过图纸与模型探索设计概念与具体处理的关系
5. 设计最终成果
时间：第7周到第8周
内容：完成最终设计文件，包括图纸与最终模型

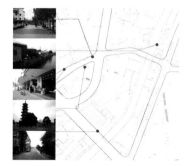
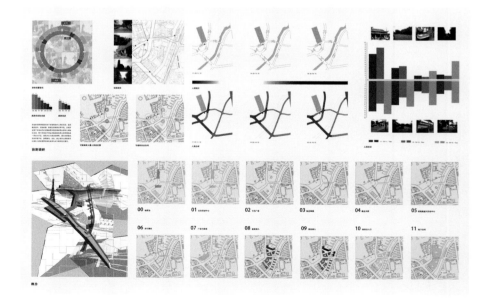

Graduate Program 1st Year
DESIGN STUDIO 2: URBAN DESIGN • FENG Lu

Type: Required Course
Study Period and Credits: 54 hours/2 credits

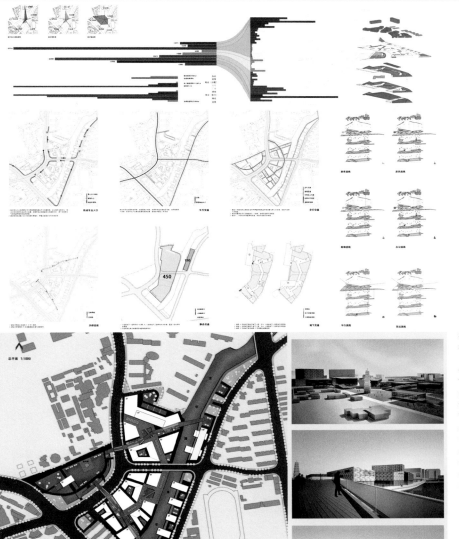

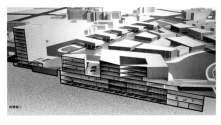

Subject: Urban Design for Renew of Longhua Temple Area, Shanghai

Course Objective
The re-built areas surrounding Longhua Temple in Shanghai involve protection of a historic site, a city update, commercial development and three-dimensional traffic organization, all of which are very complicated. These are the issues that most Chinese big cities are encountering at present. By this urban design course, students may understand meanings and ways of urban design, be aware of complicated conditions of modern cities, study ways to handle relationships between all of the elements in city when conditions are complicated, learn to define urban design guidelines and finally discover the social, political and cultural meanings in urban design.

Research Subject
Building and city, conservation and re-construction, establishment of public spaces

Design Content
The re-construction shall be done on 5 plots, which includes historic areas, commercial development sites and public green space. It may give suggestions on optimization of current planning, analyze effects of the re-construction on city surrounding areas, set and organize building functions of all plots, find out building design control principles, focus on design of historic sites to be protected, design three-dimensional traffic patterns and create the landscape spaces.

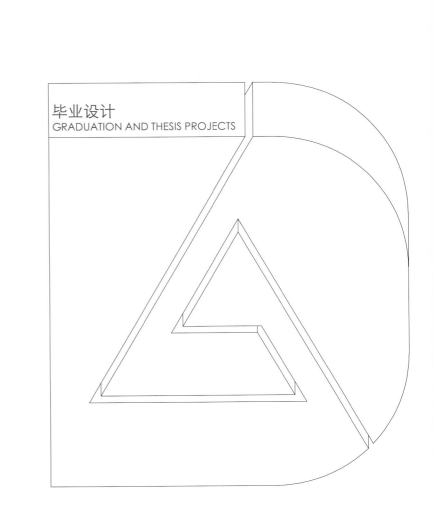

毕业设计
GRADUATION AND THESIS PROJECTS

本科四年级
本科毕业设计 • 冯金龙

课程类型：必修
学时学分：1学期

题目
1. 城市综合体建筑规划设计：南京河西江东路"紫鑫中华"三期
2. 高层旅馆建筑规划设计：南京太平南路"江苏饭店"

设计研究内容：
a. 建筑形体与城市空间关系、道路界面关系
b. 总图规划与建筑建设容量测算
c. 建筑功能构成与项目策划
d. 建筑内外交通组织
e. 建筑日照影响分析
f. 地下空间研究
g. 高层建筑幕墙构造系统与生态技术应用
h. 高层建筑结构

Subject
1. Design of Urban Complex: 3rd Phase of Zixin Zhonghua
2. Design of High-rise Hotel: Jiangsu Hotel

Design and Research Content
a. Relationship between the volume and urban space and street
b. Master plan and the estimate of development capacity
c. Function, program & project planning
d. Internal and external circulation
e. Sunlight analysis
f. Study on underground space
g. Construction system of the façade of high-rise building, application of ecological technology
h. Structure of highrise building

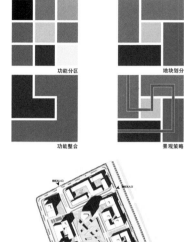

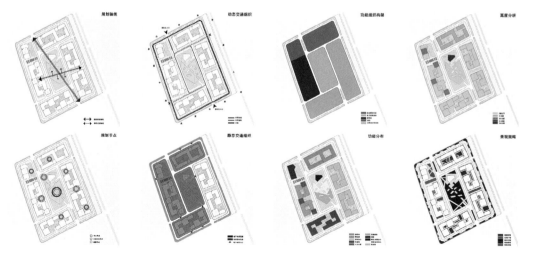

Undergraduate Program 4th Year
GRADUATION PROJECT • FENG Jinlong

Type: Required Course
Class Hours and Credits: 1 semesters

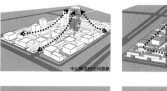

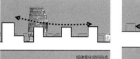

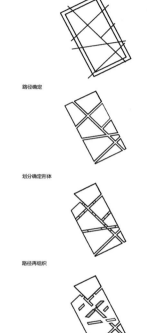

路径确定

划分确定形体

路径再组织

形体生成

STEP 1
STEP 2
STEP 3
STEP 4
STEP 5

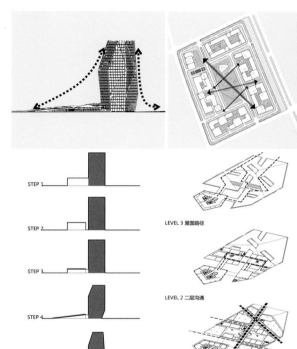

LEVEL 3 屋面路径

LEVEL 2 二层沟通

LEVEL 1 地面路径

本科四年级
本科毕业设计 • 赵辰

课程类型：必修
学时学分：1学期

题目：南京大学鼓楼校区之金陵大学老校区重整规划设计

在新形势下，南京大学鼓楼校区之金陵大学老校区，根据未来的新发展而亟须重新规划设计。"南京大学博物馆"已经是其中的一项重要发展方向，建筑与城市规划学院、医学院等相关院系的专项发展也必将进行。本课题将在大量的历史研究的基础（历史建筑测绘、文献综述）上，对南京大学"博物馆"，"建筑与城市规划学院"、"医学院"进行专项的研究与设计。

要求学生掌握建筑设计基本的技能与知识（测绘、建模、调研、分析），并能对历史建筑进行深入的设计研究（建筑结构、构造，功能策划）。

Subject: The renovation design of the old campus of former Jinling University in the Gulou Campus of NJU

The project includes the design of the Museum of NJU, the SAUP Building and the Medical School Building in Gulou Campus based on the historical research (including the survey of historical architectures and the literature reviews). Students should learn the basic knowledge and techniques of architectural design (including surveying, modeling, investigation and analysis), and study the historical architectures carefully on its structure, construction and functional program.

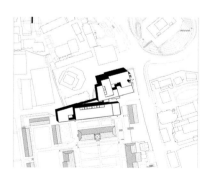
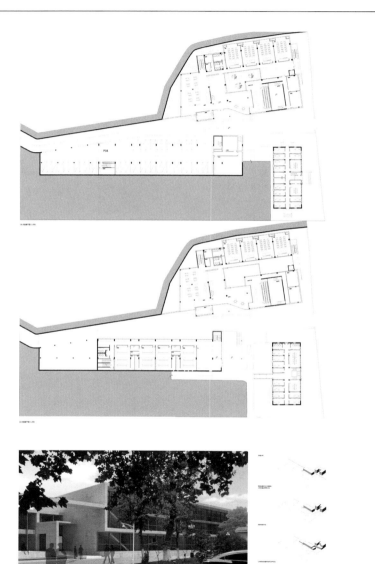
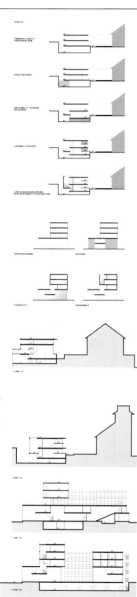

Architecture Program of SAUP, NJU 2011—2012

Undergraduate Program 4th Year
GRADUATION PROJECT • ZHAO Chen

Type: Required Course
Class Hours and Credits: 1 semesters

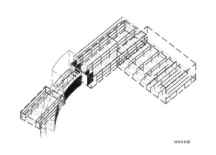
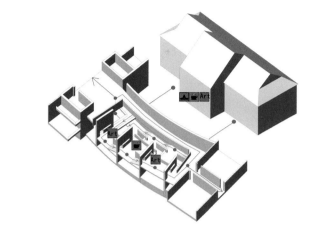

总平面 1:1000

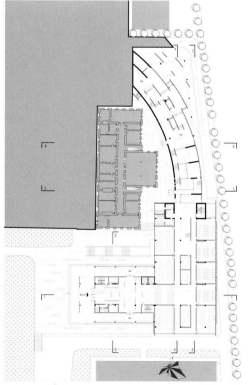
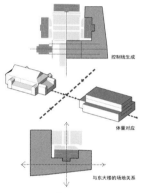

控制线生成

体量对应

与东大楼的场地关系

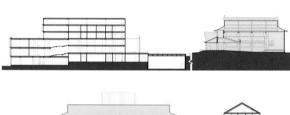

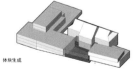

体块生成

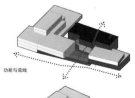

功能与流线

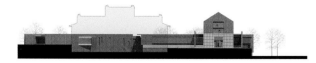

调整形态 呼应东大楼

本科四年级

本科毕业设计 • 吉国华

课程类型：必修
学时学分：1学期

题目：参数化设计与数字化建造（研究性选题）

以小型公共建筑为题探讨参数化设计和数字化建造的方法，并掌握 VB 编程基本的技能与知识。

Subject: Parametric Design and Digital Building

Research on the methodology of parametric design and digital building through the design of small public architecture, while learn the basic knowledge and technique of VB programming.

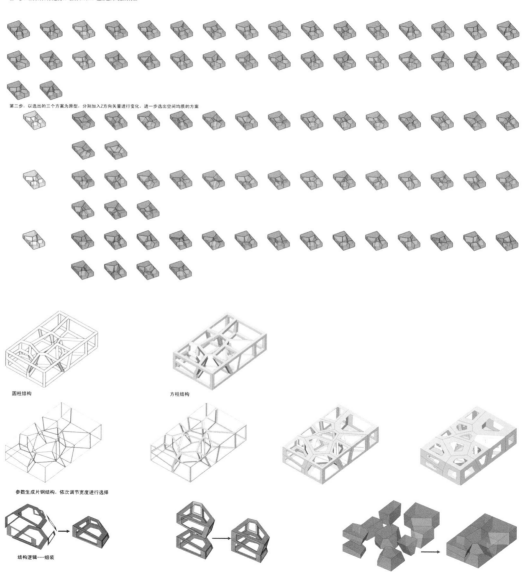

Undergraduate Program 4th Year
GRADUATION PROJECT • JI Guohua

Type: Required Course
Class Hours and Credits: 1 semesters

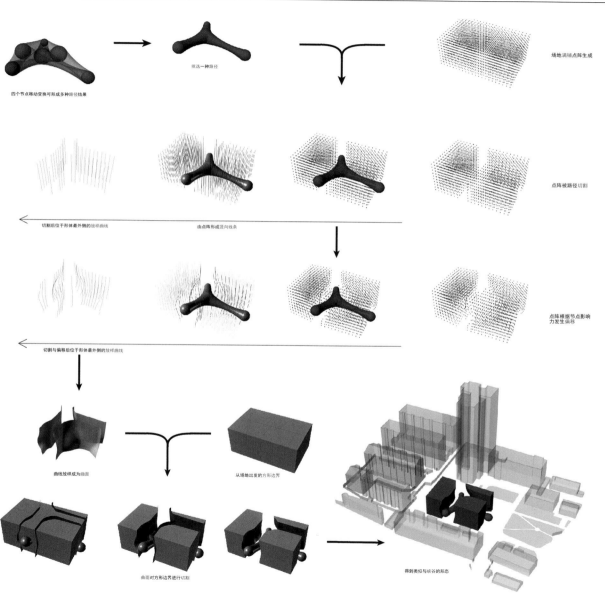

本科四年级
本科毕业设计 • 周凌

课程类型：必修
学时学分：1学期

题目：武进工业设计园区规划与单体设计

概述
　　武进工业设计园区位于常州武进新区，全区主要适用对象为国内外著名工业设计企业。项目一期规模15万m²，二期规模20万m²。目前一期规划已经确定建筑类型包括高层loft办公、SOHO公寓、独栋办公、合院住宅等。设计特殊用途的空间，既要有个性，又要有灵活性。

问题
1. 环境与场地　企业要求有风景如画的园区园区环境。
2. 策划与功能　特殊使用对象要求特殊内部空间。
3. 技术与建造　很有表现力的材料表达和技术节点。

要求
　　题目涉及城市、交通、历史等相关学科问题，训练学生在一定城市环境中进行建筑设计的各种技能，包括如何分析现状，如何处理交通与功能布局，如何处理内部空间，如何构思，如何表达，如何进行技术细化。通过学习，最终掌握分析图、平立剖图、大样图、模型等表达方法，完成建筑初步设计深度的图纸与研究报告。图纸表现方式和比例自定。

Subject: Wujin Industrial Design Park Planning and Monomer design

Introduction
Wujin Industrial Design Park locates in Wujin New Area of Changzhou, the region mainly intended for the famous industrial design firms. The first phase includes a scale of 150,000 m², and the second 200,000 m². A planning building types have been identified, including high-level loft office, SOHO apartments, single-family office, residential courtyard. Design special purpose space with personality, as well as flexibility.

Problem
1. Context and Site
2. Program and Function
3. Technique and Construction

Mission Requirements
The subject concerns problems of urbanism, transportation, history and relational disciplines. It trains students' architectural design skills in certain urban environment, including how to analyze the situation, how to deal with traffic and functional layout, how to deal with the internal space, how to generate an idea, how to express, and how to design the detail. Students should grasp the express methods using analysis charts, plans, elevations, sections, detail drawings and models, then finish the design drawings and research reports.

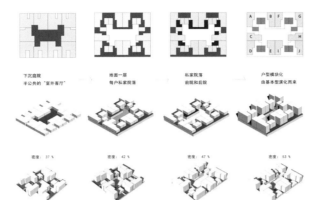

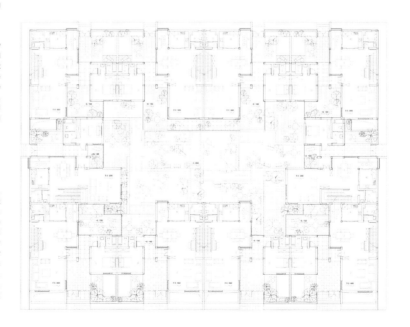

Undergraduate Program 4th Year
GRADUATION PROJECT • ZHOU Ling

Type: Required Course
Class Hours and Credits: 1 semesters

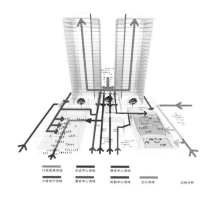
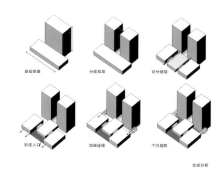
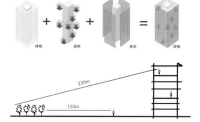
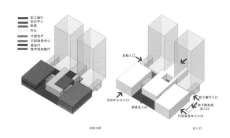
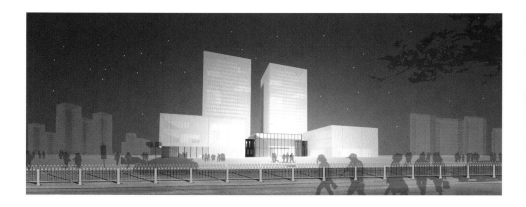

本科四年级

本科毕业设计 • 傅筱 秦孟昊

课程类型：必修
学时学分：1学期

题目：从节能角度启动的建筑——南京紫东国际创意产业园 A7 栋办公楼设计

　　南京紫东国际创意产业园是一个正在建设中的江苏省节能示范产业园区。如何将节能作为一项基本要素介入建筑设计之初，并与建筑的使用功能、场地、艺术、空间紧密地融合在一起，是值得研究的课题。建筑规模约 10000 m²，多层。
　　要求学生掌握建筑设计基本的技能与知识，并能够熟练使用节能分析软件进行设计推动。

Subject: Architecture Start from Energy Efficiency — Design of the A7 Office Building in Nanjing Zidong International Creative Industry Park

Zidong International Creative Industry Park of Nanjing is a construction of energy-saving demonstration Industry Park in Jiangsu province. How to make the energy-saving as one of the basic intentions involved in the beginning of architectural design while integrating closely with function, site, art and space is a subject worthy to study. Construction area is about 10,000 m² with multi layers.
Students are required to learn the basic knowledge and methods of architectural design push the design forward using the energy-saving analysis softwares.

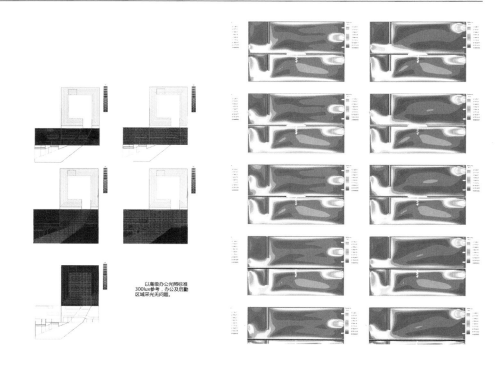

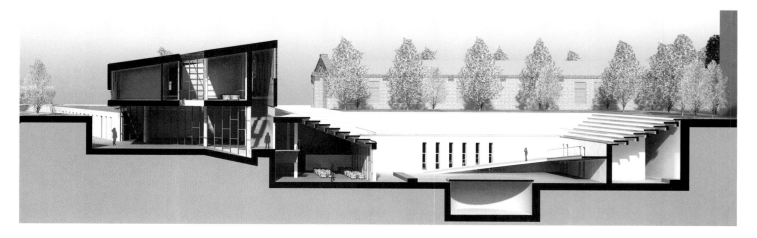

Architecture Program of SAUP, NJU 2011—2012

Undergraduate Program 4th Year

GRADUATION PROJECT • FU Xiao, QING MengHao

Type: Required Course
Class Hours and Credits: 1 semesters

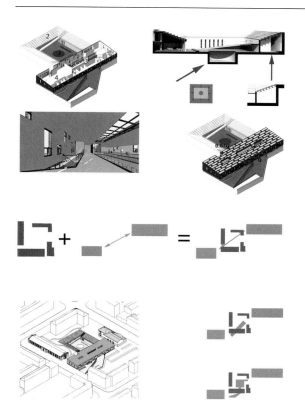

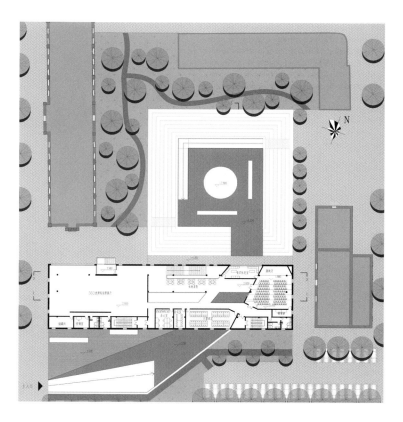

本科四年级

本科毕业设计 • 华晓宁

课程类型：必修
学时学分：1学期

题目：新疆克孜勒苏柯尔克孜自治州"三馆"（博物馆、图书馆、文化馆）方案设计

随着各地社会经济发展水平的提高，对于文化建筑的需求日益彰显。此类建筑对于弘扬地域文化、丰富市民的精神生活具有较为重要的意义。本课题将研究在特定的文化背景下，设计能够承载和表达地域特质的文化综合体建筑。

要求学生掌握建筑设计基本的技能与知识，并能够对特定的文化传统和地域特征进行深入研究，探索将其以特定的建筑语汇进行表达的策略。

Subject: Design of the Museum, Library and Cultural Center at Kergez Autonomous Prefecture of Kizilsu

Design a cultural complex which can contain and represent the regional characters under the background of specific culture. Students should learn the basic knowledge and technique of architectural design, study carefully on the specific cultural traditions and regional characters, research the expressional strategies in specific architectural language.

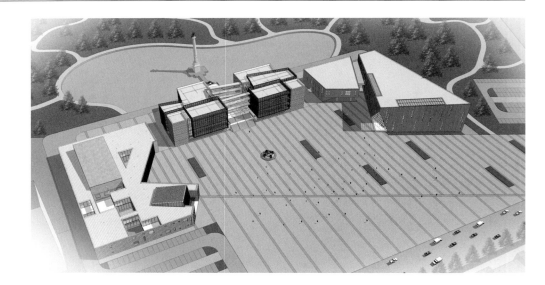

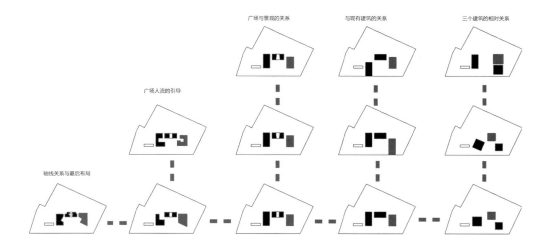

Undergraduate Program 4th Year
GRADUATION PROJECT • HUA Xiaoning

Type: Required Course
Class Hours and Credits: 1 semesters

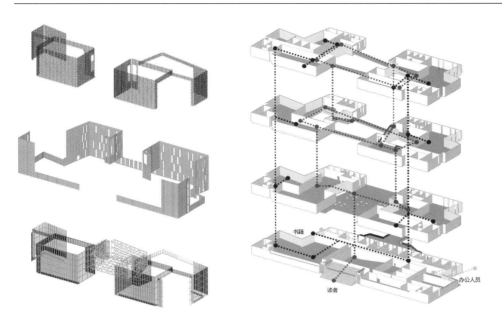

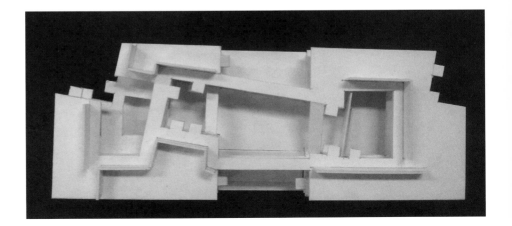

研究生二年级
专业硕士毕业设计：城市设计研究

课程类型：必修
学时学分：1学期

题目：上海徐汇中漕新村143地块文化街区设计
指导教师：吉国华 教授

本项目位于上海徐汇区中心区，内环高架与沪闵路高架的相交处，西临中漕路，南近中山西路，东靠漕溪北路，北临柿子湾街，地块面积26221m²。附近有徐家汇商圈、上海体育场、宜家等大型配套设施。规划商业办公文化用地中，商业、办公和文化娱乐三类的建筑面积分别占地上建筑总面积的40％、55％和5％。

该项目位于城市繁华地带，环境复杂，在规划及建筑设计中应注重与城市文脉的关系。除了创造一个全新的、高品质的商业办公文化中心，还应融合城市文化，赋予人文内涵。

Subject: Design of the Culture Block at 143 plots of Zhongcao Village in Xuhui District of Shanghai
Advisor: Professor JI Guohua

The project is located in the center area of Xuhui District in Shanghai, the inner ring viaduct and Humin Road elevated intersection, west of North Caojing Road, near Zhongshan Road. The land area is 26,221m², with the shopping district of Xujiahui, Shanghai stadium, IKEA and other large facilities nearby. The gross floors area(GFA) of commercial, office and entertainment buildings should cover 40%, 55% and 5% of the total GFA on the land for commercial, official and cultural using.

The project is located in the downtown area with complex environment. The relationship with urban context should be payed much more attention in urban planning and architectural design. In addition to creating a new high-quality commercial, official and cultural center, it should be integrated with urban culture and humanities connotation.

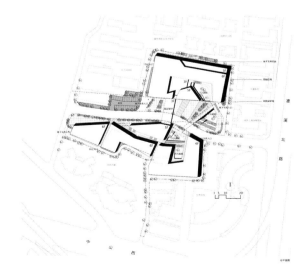

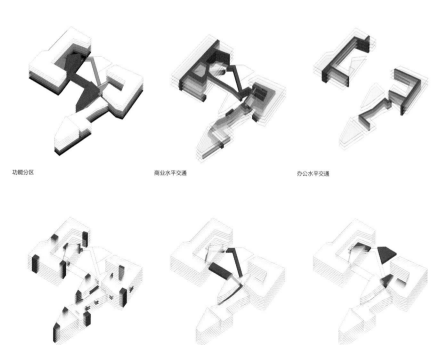

功能分区　　　商业水平交通　　　办公水平交通

垂直交通　　　南北区联系　　　下沉庭院

Architecture Program of SAUP, NJU 2011—2012

Graduate Program 2nd Year
THESIS PROJECT: URBAN DESIGN

Type: Required Course
Study period and credits: 1 semester

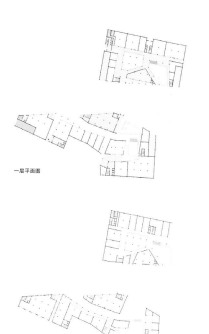

一层平面图

二层平面图

总平面形态生成

设计行走路径,南侧商业外街,北侧为室内街道 → 天桥联系,布置展览空间 → 下沉庭院联系,布置负一层商业 → 化整为零,生成次街空间

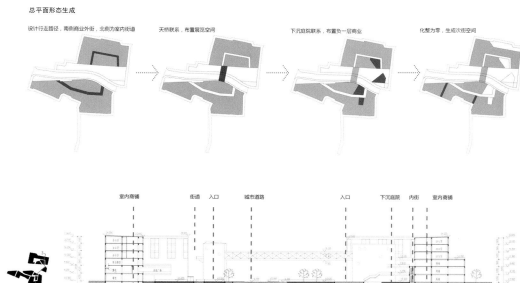

室内商铺 街道 入口 城市道路 入口 下沉庭院 内街 室内商铺

研究生二年级
专业硕士毕业设计：城市设计研究

课程类型：必修
学时学分：1学期

题目：上海天目中路高架下空间重构设计
指导教师：华晓宁 副教授

本课题以当代大都市基础设施空间利用为研究背景，对上海市高架桥下空间利用现状进行大规模的调研，选择某一特定地块作为毕业设计的基地，进行功能、空间的重构。

高架下空间的重构，主要是变高架下废弃空间为有用空间，成为城市公共空间的一部分，空间模式有别于一般的城市空间。本课题以高架下人的行为活动模式研究作为出发点，进行桥下空间的重新设计和利用。

Subject: The Renovation of Urban Space under Middle Tianmu Road Viaducts in Shanghai
Advisor: Associate Professor HUA Xiaoning

This project investigates the usage of urban space under the viaducts in Shanghai under the background of reusing infrastructure space in contemporary metropolis, then chooses one certain site to reconstruct the function and space.
The renovation of urban space under the viaducts is to transform the waste space into active urban public space. The mode of this space is different from normal urban space. This project redesigns and reuses the space under the viaduct from the start point of the mode of human behavior.

基地

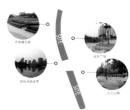
现状

问题与策略
城市层面

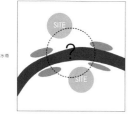

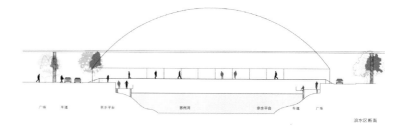
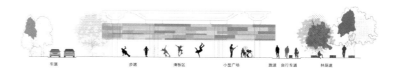

Graduate Program 2nd Year
THESIS PROJECT: URBAN DESIGN

Type: Required Course
Study period and credits: 1 semester

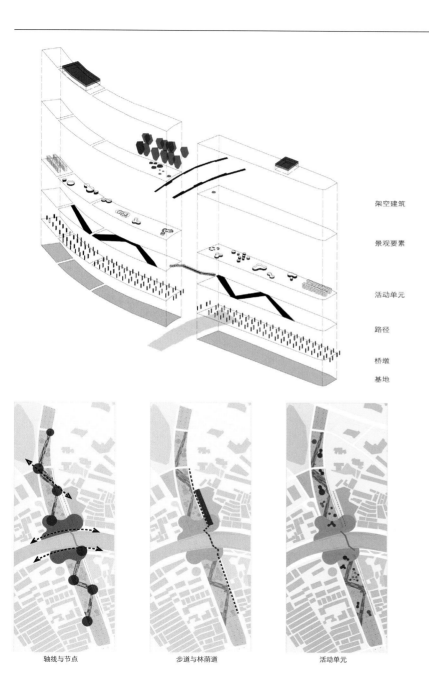

架空建筑
景观要素
活动单元
路径
桥墩
基地

轴线与节点　　步道与林荫道　　活动单元

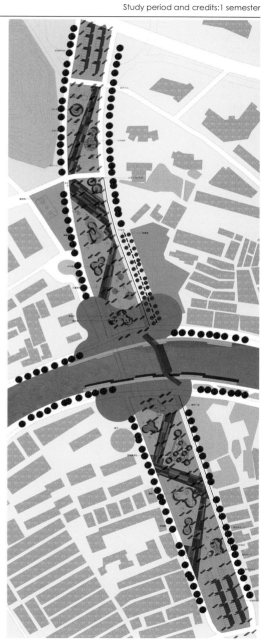

研究生二年级
专业硕士毕业设计：公共建筑设计

课程类型：必修
学时学分：1学期

题目：南京大学科学园文化创意产业园1 - 4号楼
导师：张雷 教授

南京大学科学园文化创意产业园工程项目位于南京市栖霞区仙林新市区白象片区内，元化路以西，九乡河西路以东，北临纬地路，南临南京大学仙林校区。基地为山凹处，现场有一个天然的半圆形缺口，地势南高北低，南侧为高差较大的山地。建筑主要功能为科研办公。总用地面积为297961 m²，建筑物控制高度不超过24 m，容积率≤1.5，建筑密度≤30%。

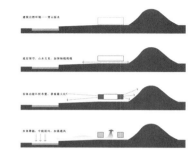
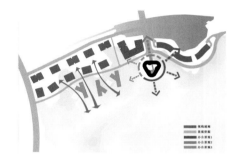

Subject: 1# - 4# Buildings of Cultural and Creative Industry Park in Nanjing University Science Park
Advisor: Professor ZHANG Lei

The Cultural and Creative Industry Park in Nanjing University Science Park is located in the White Elephant Chip Area of Xianling New Town in Qixia District of Nanjing, north to Weidi Road, south to Xianlin Campus of Nanjing University, east to the West Jiuxiang Road and west to Yuanhua Road. The site is a hollow with a semi-circular gap. The land lean from south to north, while the south part is the mountain with big altitude difference. The function of new building is office.

Total land area: 297,961 m². Height of buildings: ≤ 24m.
Floor area ratio: ≤1.5. Building density: ≤ 30%.

Graduate Program 2nd Year
THESIS PROJECT: PUBLIC BUILDING DESIGN

Type: Required Course
Study period and credits: 1 semester

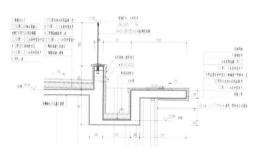
1号楼墙身大样图

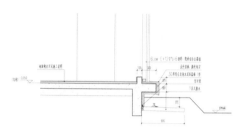

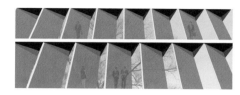

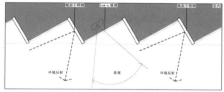
1号楼立面

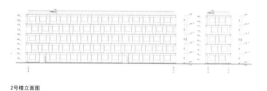
2号楼立面图

1号楼立面图

2号楼立面

2号楼平面图

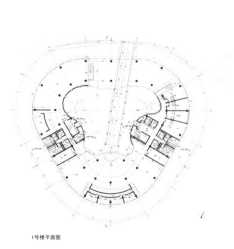
1号楼平面图

研究生二年级
专业硕士毕业设计：公共建筑设计

课程类型：必修
学时学分：1学期

题目：泰州市泰州大学共享区图书馆建筑设计
导师：周凌 副教授

随着高校规模的扩大，校园改扩建建设成为近十年的发展潮流。随着科学的进步和技术的发展，高校图书馆建筑也在随之创新和完善，并出现了新的跨越。本设计通过对国内外高校图书馆建筑的发展状况研究，解析高校典型图书馆的基本布局、组织手法等特点，围绕合并后的泰州市泰州大学图书馆建筑单体设计的展开、进行过程，回应新时期高校图书馆的设计趋势，并对目前图书馆尚待改进的地方归纳总结，在实际设计过程中扬长避短。

本方案整体构思为"方院+轴线"模式。

Subject: Design of The Library in Taizhou University
Advisor: Associate Professor ZHOU Ling

With the expansion of universities in China, the extension and reconstruction of campus become the trends of development in recent ten years. Meanwhile, the university libraries innovate, improve and even make new progress with the development of science and technology. This project studies the development of domestic and foreign university libraries, analysis the basic spatial layout and organizing methods of typical university libraries, responds to the design trends in the new era with the design of the Library in Taizhou University after the consolidation. It also summarizes the primary problems in contemporary design of libraries which should be improved in the future.

Main idea of the project is the mode of "Court + Axis".

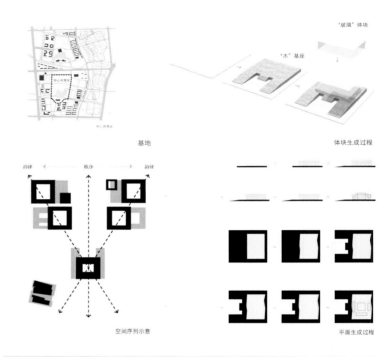

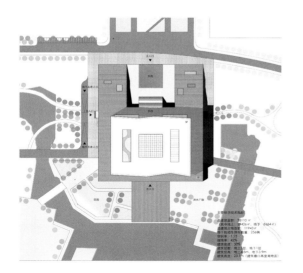

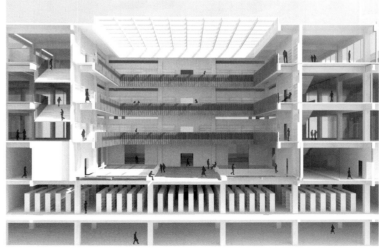

Graduate Program 2nd Year
THESIS PROJECT: PUBLIC BUILDING DESIGN

Type: Required Course
Study period and credits:1 semester

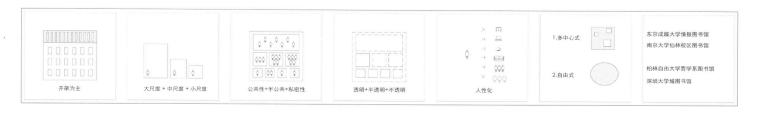

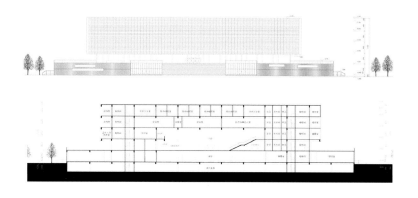

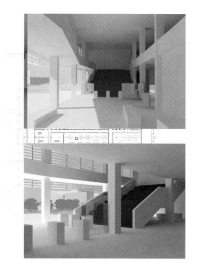

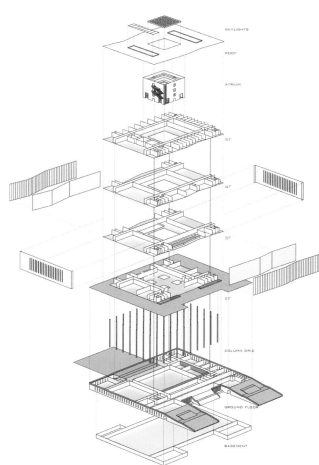

研究生二年级
专业硕士毕业设计：建构设计研究

课程类型：必修
学时学分：1学期

题目：东太湖大厦建筑设计与构造研究
导师：周凌 副教授

本项目基地位于苏州吴江市新开发的滨湖新城区，项目占地面积约199亩，规划容积率为0.37，主题功能为办公、城市规划展览、餐厅宴会、会议中心、行政服务中心等。功能定性为滨湖新城区的管委会大楼。

本次设计的内容是方案的深化设计和构造研究，主要为立面幕墙和百叶的构造原理研究，用百叶塑造立面。塔楼的东西方向可以通过竖向百叶来达到遮阳效果，在塔楼的南向，由于竖向遮阳效果不太理想，所以采用横向百叶加竖向百叶综合遮阳的方式。对幕墙、百叶构造原理研究后，针对建筑方案进行构造、材料、大样的深化设计。还综合考虑构件尺度合理、方便性，对施工方式进行深入研究。

Subject: Architectural Design and Tectonic Study of East Taihu Tower
Advisor: Associate Professor ZHOU Ling

The project is located in the Lakeside New City of Wujiang. The land area is about 199 mu. The plot ratio is 0.37. The building is for the Administrative Committee of Lakeside New City with functions of offices, exhiibition of urban planning, restaurants, conference centers, administrative service center, etc.

Contents of this project include the development of design and the tectonic study focusing on the details of the skin and louver boards on the facade. The facade is mainly characterized by louver boards. The form and configuration of louver boards follow the sunlight. The details, materials and operational methods are also carefully studied.

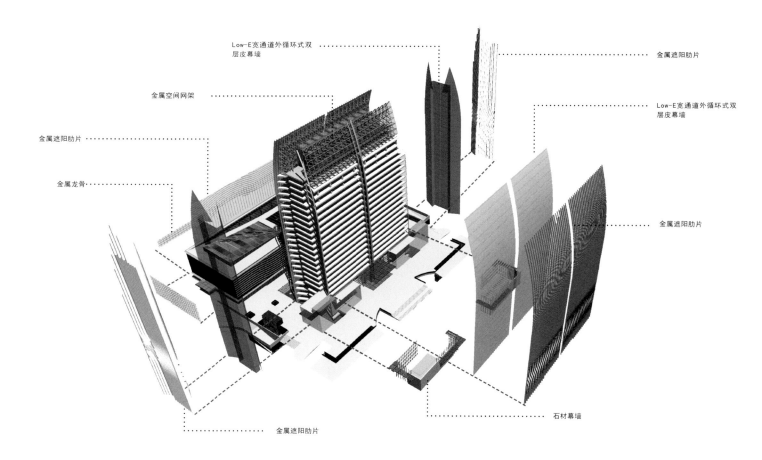

Architecture Program of SAUP, NJU 2011—2012

Graduate Program 2nd Year
THESIS PROJECT: CONSTRUCTION DESIGN

Type: Required Course
Study period and credits: 1 semester

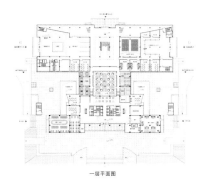
一层平面图

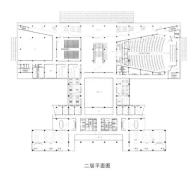
二层平面图

标准层平面图

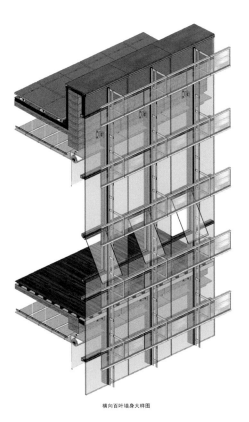
横向百叶墙身大样图

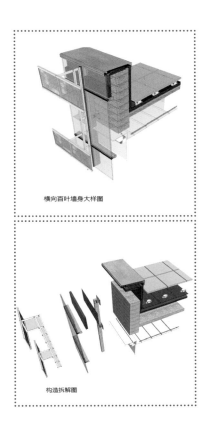
横向百叶墙身大样图

构造拆解图

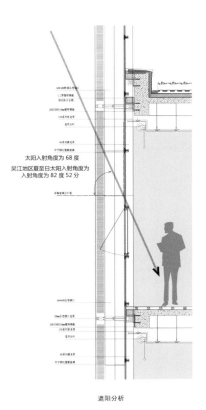
遮阳分析

研究生二年级
专业硕士毕业设计：建构设计研究

课程类型：必修
学时学分：1学期

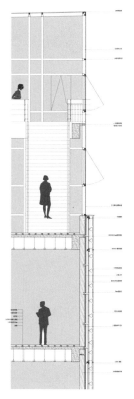

位置4：大样图

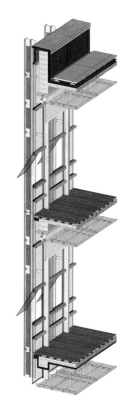

竖向百叶墙身大样

横向百叶墙身大样

石材幕墙加玻璃幕墙大样图

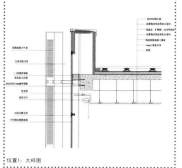

位置1：大样图

位置2：大样图

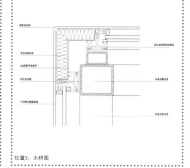

位置3：大样图

剖面位置示意图

Graduate Program 2nd Year
THESIS PROJECT: CONSTRUCTION DESIGN

Type: Required Course
Study period and credits:1 semester

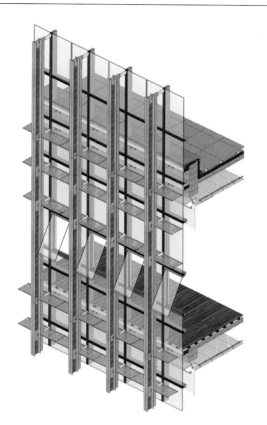

竖向百叶加横向遮阳大样图

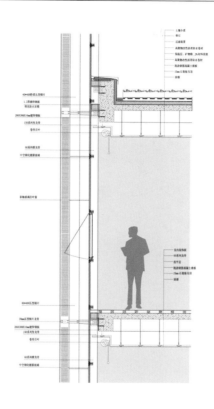

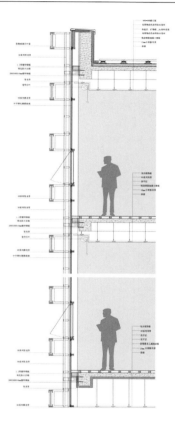

横向百叶墙身大样图

玻璃幕墙角部大样图

横向百叶大样图

竖向百叶大样图

石材幕墙构造节点大样图

本科二年级

专业硕士毕业设计：绿色建筑设计

课程类型：必修
学时学分：1学期

题目：泰州北核心区城市设计和基于噪声环境下的建筑设计
导师：丁沃沃 教授

在建筑设计中，针对拟设计地块的具体影响因素：噪声，做出进一步的分析总结。声环境是建筑环境的重要组成部分。人们学习、工作和休息都需要安静的建筑环境。噪声的危害逐渐被人们重视，这充分体现了控制声环境的重要。如何控制高层建筑的声环境影响已是一个亟待解决的问题。尝试运用经认证的标准化声学软件CadnaA 对高层建筑的声环境进行模拟，了解高层建筑声环境的分布规律，提出建筑设计上的策略。具体方法是：从最初形体生成，到各功能区域组织，再到高层建筑的表皮设计，应用各种对抗噪音的建筑策略，使得高层建筑在声环境较差的区域也可以实现自然通风，从而减少能耗。

Subject: Urban Design of Taizhou North Core Area and Architectural Design in the Noise Environment
Advisor: Professor DING Wowo

In the architectural design, further analysis and summaries were applied for the proposed site specific factor: noise. Noise environment is an important element of building environment. A quiet enviroment is required for studying,working and resting. Much more attentions have been paid on the detriment of noise, which embody the importance of acoustics environment control. How to control the acoustics environment of high-rise building has been a urgent problem.The project tries to use the authorized and standardized acoustics soft CadnaA to simulate the acoustics environment of high-rise building, studies the noise level distribute data out of high-rise buildings and proposes design strategies.The specific methods is to apply various strategies against noise from the initial form generating to the organization of each functional zones and then the facade design of high-rise buildings, so that the high-rise buildings can get natural ventilation even in the area of serious acoustics environment, thereby reduce the energy consumption.

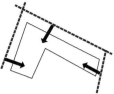

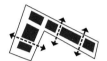

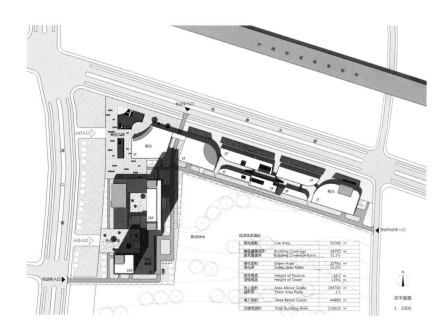

THESIS PROJECT: GREEN BUILDING DESIGN

Type: Required Course
Study period and credits: 1 semester

塔楼布局

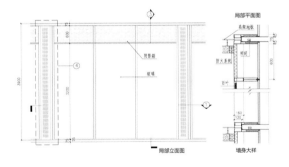

研究生二年级
专业硕士毕业设计：绿色建筑设计

课程类型：必修
学时学分：1学期

题目：戊己庚楼自然通风改造设计
导师：傅筱 副教授

本设计以南京大学戊己庚楼改造为研究对象，现有建筑为文物保护建筑，具有保存价值，但其空间和舒适性已经很难满足使用要求，原建筑为东西朝向，且周围建筑密集，建筑通风条件尤其需要改善。因此，在本设计中重点考虑了对自然通风的设计。而且在当今，能源危机和可持续发展的大背景下建筑设计中更应该考虑被动式设计而不是单纯依靠机械。着重分析了CFD技术在建筑设计中的应用，通过分析建筑所处地域气候后，针对建筑自身的特点，选择合适的自然通风策略，并借助CFD进行模拟分析，从宏观和微观上反映出自然通风的效果。通过模拟分析结果对建筑设计方案进行调整，保证通风策略的有效性和可靠性。设计分别从场地、空间和细部三方面对建筑进行了模拟分析后的改造设计，以期总结归纳出一些对自然通风设计有参考价值的设计策略。

Subject: Design of Natural Ventilation in the Renovation of E, F, G Buildings
Advisor: Associate Professor. FU Xiao

Nanjing University, E & F G Building Renovation, the design for the study, analyzed the application of CFD in building design through the analysis of the building in which the regional climate, building its own characteristics, select the appropriate natural ventilation strategy, and with CFD simulation analysis to reflect the effect of natural ventilation from the macro and micro. Simulation results on the architectural design be adjusted to ensure the validity and reliability of the ventilation strategy. Designed separately from the face of building space, space and detail tripartite design of simulation analysis after the transformation.

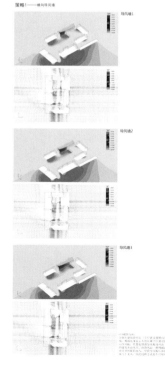

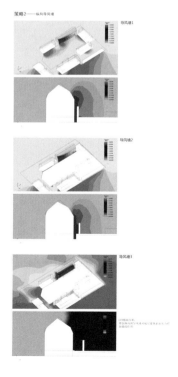

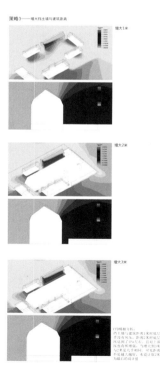

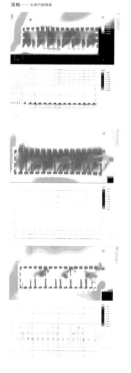

Architecture Program of SAUP, NJU 2011—2012

Graduate Program 2nd Year
THESIS PROJECT: GREEN BUILDING DESIGN

Type: Required Course
Study period and credits: 1 semester

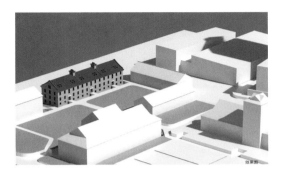

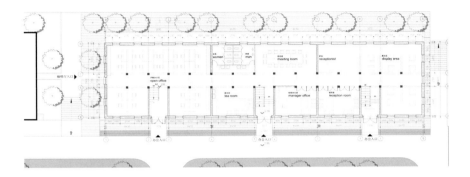

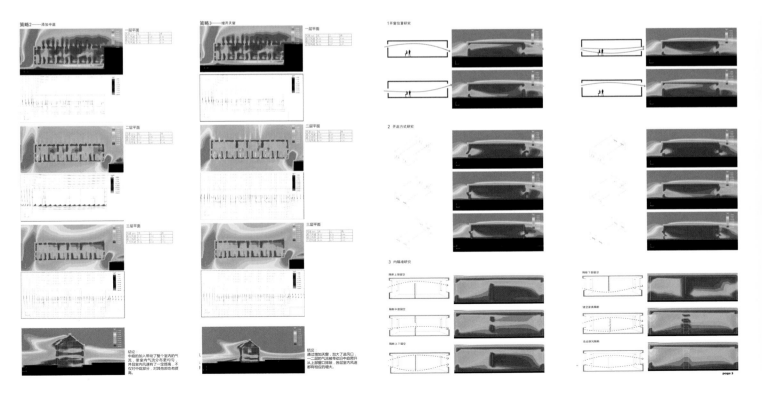

研究生二年级
专业硕士毕业设计：建筑遗产再生

课程类型：必修
学时学分：1学期

题目：传统空间意象需求下装折界面的现代演绎
导师：萧红颜 副教授

空间的划分问题一直是设计的基本问题，传统空间划分之用的装折，展现了其灵活性及对可变空间的塑造性。作为小木作的装折对传统空间意象的营造起着重要作用。在当今传统空间意象需求下的新型建造体系中，如何通过装折的处理达到功能需求与空间划分的一致性，同时展现空间的可变性及灵活性值得研究。本项目主要研究传统建筑中的装折在现代空间及材料中的运用和发展，以求通过装折设计实现既传统又现代的空间意象。

本设计选取"蒋寿山地块体验式精品酒店"的大堂进行建筑及装折设计。在较为完整的建筑设计基础上，从装折的位置、类型、尺度、构造等方面对之前所做的相关研究进行设计运用。

Subject: The Modern Deduction of Zhuangzhe Interface with Traditional Spacial Image
Advisor: Associate Professor XIAO Hongyan

The division of space is one of the basic problems of design. The traditional space partition Zhuangzhe has shown out the flexibility and capability of shaping variable space. As one kind of joinery works, Zhuangzhe plays an important role in creating the traditional spacial image. In the contemporary new construction system under the needs of the traditional spacial image, it is worth to study how to use Zhuangzhe to achieve the consistency of function requirements and space partitions while revealing the variability and flexibility of space. The project studies the application and development of traditional Zhuangzhe in modern space and materials, thereby achieve spacial images integrating the tradition and modernity.

The project takes the lobby of Jiangshoushan Experiential Boutique Hotel as an example to do the Zhuangzhe design, applying the results of former researches on the aspects of position, type, scale and tectonic of Zhangzhe based on the holonomic architectural design.

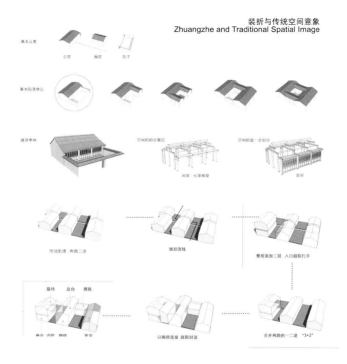

装折与传统空间意象
Zhuangzhe and Traditional Spatial Image

Graduate Program 2nd Year
THESIS PROJECT: BUILDING'S REGENERATION

Type: Required Course
Study period and credits: 1 semester

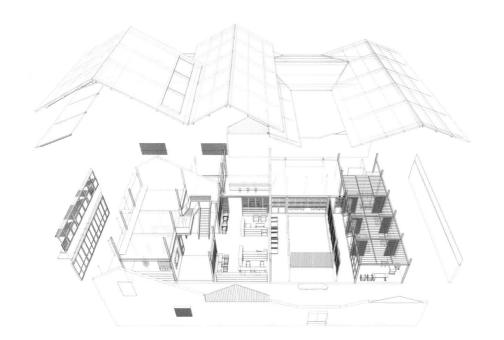

装折位置初定 Position of Zhuangzhe

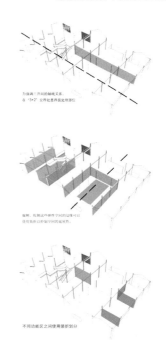

研究生二年级
专业硕士毕业设计：建筑遗产再生

课程类型：必修
学时学分：1学期

架构生成 Structure Generation

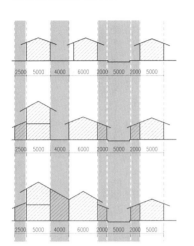

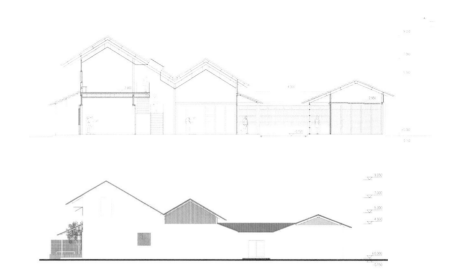

功能流线分析

原规划流线
主要流线
辅助流线

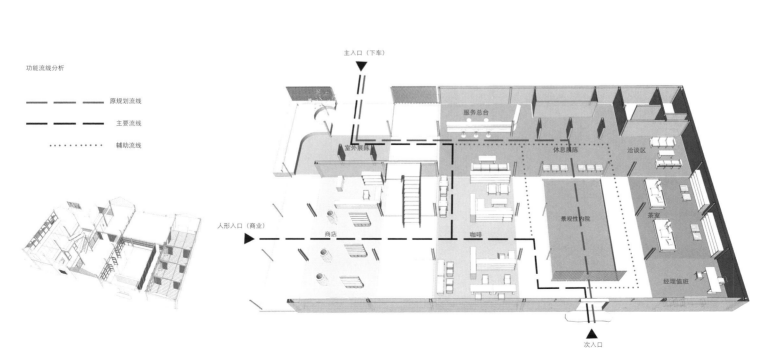

Architecture Program of SAUP, NJU 2011—2012

Graduate Program 2nd Year
THESIS PROJECT: BUILDING'S REGENERATION

Type: Required Course
Study period and credits: 1 semester

装折界面划分与尺度　Division and Scale of Zhuangzhe Interface

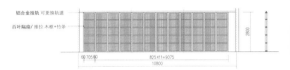

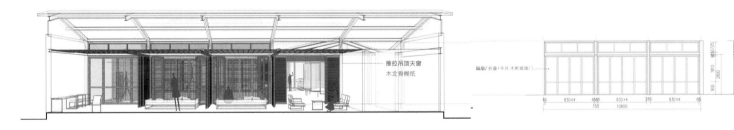

设计工作坊
DESIGN WORKSHOPS

本科二年级
古建筑测绘 • 赵辰 萧红颜

课程类型：必修
学时学分：36学时/2学分

课程介绍

1. 教学内容

分为"测"与"绘"两个教学环节，通过"测"与"绘"两个阶段进行相关的专业训练。"测"：由实地实物的尺寸数据的观测量取；"绘"：根据测量数据与草图进行处理、整理最终绘制出完备的测绘图纸及报告。区别于以往单纯图纸绘制方式的测绘教学，要求学生通过问题研究的思维与视野，思考讨论相关专业问题。个人独立答辩题目为："测"与"绘"之感受——对中国传统建筑空间的初步理解。

2. 教学安排

分为现场测绘、整理图纸与报告两个环节。

其一：现场测绘。总体分析记录建筑群所在的环境特点和总体空间特征；逐一测绘单体建筑的结构样式、构件尺寸及特殊做法。

其二：整理图纸。第一步骤，整理测绘图纸，建筑测绘图纸的内容包括总平面、总剖面以及单体建筑的各层平面、剖面、立面及相关大样，并绘制三维建筑模型。第二步骤，编写测绘报告，测绘报告包括对测绘图纸不易表达内容的说明及对相关问题的思考。

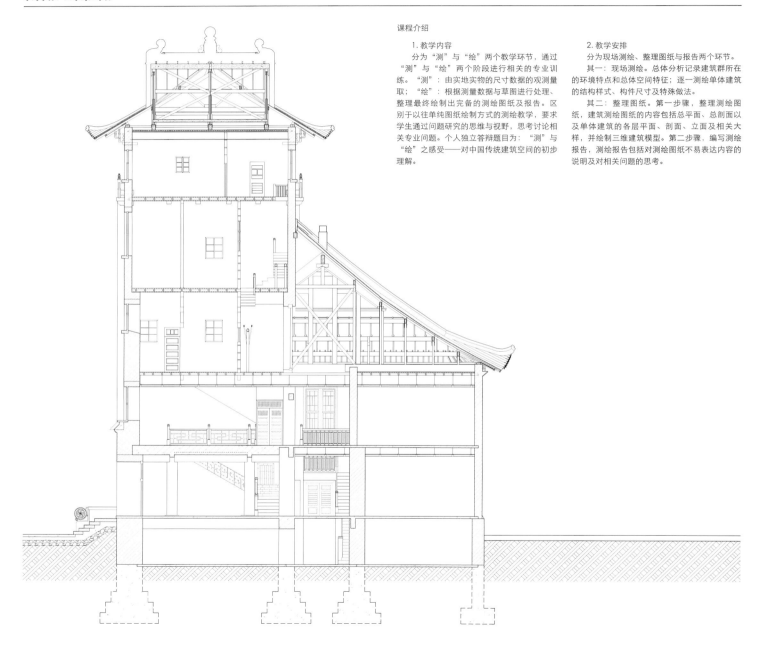

Undergraduate Program 2nd Year

ANCIENT BUIDING SURVEY AND DRAWING • ZHAO Chen, XIAO Hongyan

Type: Required Course
Study Period and Credits: 36 hours/2 credits

Course Introduction

1. Teaching Content

Key teaching points include "surveying" and "mapping". Carry out related professional training by two stages – "surveying" and "mapping". "Surveying": observe and take the dimensional data of physical objects from physical sites. "Mapping": based on the surveyed data and draft drawings, process, reorganize and finally draw complete survey maps and reports.

Different from the past surveying and mapping teaching in simple methods of drawing maps, this course requires students to think and discuss related professional problems through the thinking and vision of problem research. The title for independent defense is: Feeling of "surveying" and "mapping" – preliminary understanding of China's traditional building space.

2. Teaching Arrangement

It includes two steps, i.e. on-site surveying & mapping and drawings & reports reorganization.

(1) On-site surveying and mapping. Overall analysis and recording of the environmental characteristics and the overall spatial characteristics of the place where the ancient building complex is located; survey and map the structure patterns, component sizes and special practices of single buildings on a one-by-one basis.

(2) Drawings reorganization. Step 1. reorganize the surveying and mapping drawings. The surveying and mapping drawings of ancient buildings include master plans, master profiles, plans, profiles, elevation and the related details of all floors of a single building. And draw up a 3-dimensional building model. Step 2. prepare a surveying and mapping report, including explanations of content that is difficult to express on the surveying and mapping drawings and the thinking of the related problems.

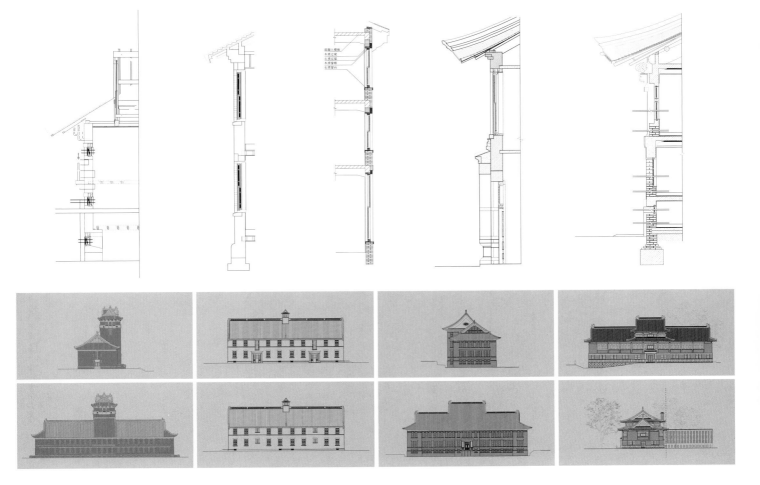

研究生一年级

中国建构（木构）文化研究 • 赵辰　冯金龙

课程类型：必修
学时学分：18学时/1学分

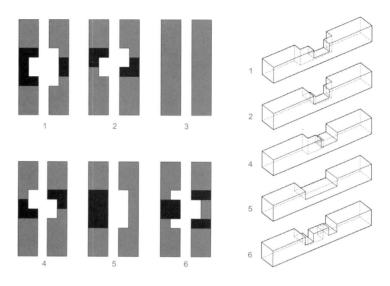

STUDIES IN WOODEN TECTONIC CULTURE/2012 STUDENT CONSTRUCTION FESTIVAL IN XIANLIN CAMPUS, NJU　Studio-1 From Material to Construction: Liu Mu Tong Gen

工作室-1 从材料到构造："六木同根"

以木为材料的建构文化是各种文明中的基本成分，中国的木建构文化更是深厚而丰富。在全球可持续发展要求之下，木建构文化必须得到重新的认识和评价。对于中国建筑文化来说，更具有文化传统再认识和再发展的意义。

文化的个性和差异并不仅仅存在于建筑的形态之中，更体现在建构的过程之中。"木建构文化研究"从木材的基本材料特性出发，研究木材的连接形成木建构的形态，探索以此构成各种功能目的的营造物。从木构的结构造型可以看出，构件的分解、组合有可能更充分地发挥木材的材质特性，从而突破传统木构的高度与跨度之限制。通过不同文化的木建构对比，有助于建筑师理解木构文化的意义。

The tectonic culture based upon wooden material is the fundamental part of various civilizations in the world, while the wooden tectonic culture in China has been relatively richer. The wooden tectonic culture must be redefined and revaluated along with the global sustainable development. For Chinese architectural culture, it would be meaningful to redefine and redevelop cultural tradition in China.
The cultural significance and differentiation should not only be shown in the form of architecture, but also expressed in the process of tectonics. "Studies in Wooden Tectonic Culture" started from the basic element of wooden material, is to study about the wooden tectonic components through the connection of wooden pieces, and to explore various functioned constructions. The recognition could be reached based upon a review of the development of wooden structure forms in the world, the characteristic of timber material would be more liberated just with the dissembling and assembling of elements, therefore it breaks through the limitation of traditional wooden structure. It would be great help for architects to understand the meaning of wooden culture, just with comparative studies between different wooden cultures.

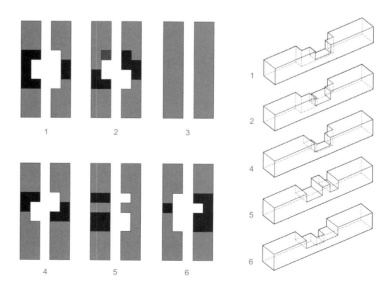

Graduate Program 1st Year

STUDIES IN CHINESE WOODEN TECTONIC CULTURE • ZHAO Chen, FENG Jinlong

Type: Required Course
Study Period and Credits: 18 hours/1 credits

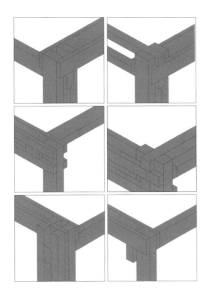

工作室-2 从构造到结构单元：木构框架
Studio-2 From Construction to Structure Unit: Wooden Frame

STUDIES IN WOODEN TECTONIC CULTURE/2012 STUDENT CONSTRUCTION FESTIVAL IN XIANLIN CAMPUS, NJU

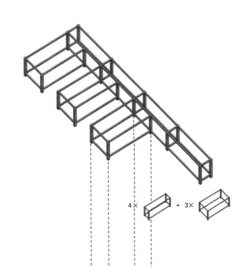

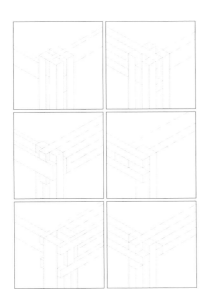

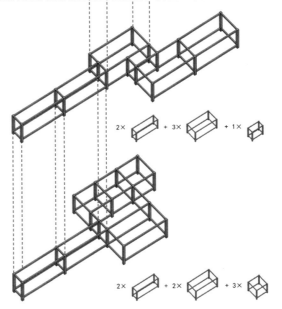

研究生一年级

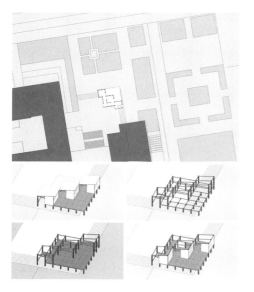 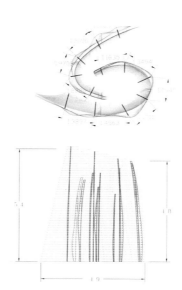 

STUDIES IN WOODEN TECTONIC CULTURE/2012 STUDENT CONSTRUCTION FESTIVAL IN XIANLIN CAMPUS, NJU 工作室-3 根据场地/功能的基本单元发展 Studio-3 Development of Basic Units According to the Site

设计一
DESIGN I

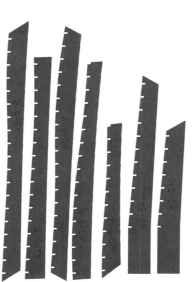

设计二
DESIGN II

设计三
DESIGN III

工作室-4 建造实验方案的发展
Studio-4 Development the Project of Construction Experiment   STUDIES IN WOODEN TECTONIC CULTURE/2012 STUDENT CONSTRUCTION FESTIVAL IN XIANLIN CAMPUS, NJU

设计一
DESIGN I

设计二
DESIGN II

设计三
DESIGN III

研究生一年级

数字建筑设计 • 吉国华

课程类型：选修
学时学分：36学时/2学分

题目：系列自行车棚

采用数字设计技术，在南京大学鼓楼或仙林校区改建或新建一批形态新颖的自行车棚，使之成为校园景观的积极因素。

设计主题

"无形"生"有形"

将看似"无形"的因素，例如风、力、音乐旋律、数列等，通过某种方式转换为建筑的形式。

设计合作

每个设计工作组由两位同学组成。

设计平台

RhinoScript/Grasshopper

Subject: A Series of Bike Sheds

The designers would use parametric techniques to rebuild or build a series of bike sheds with novelty formation in Gulou or Xianlin Campus of Nanjing University. The bike sheds should be a positive factor of campus landscapes.

Theme of the Design

From "invisible" to "visible"
The designers should use parametric techniques to transform invisible factors, such as wind, music, sequence of number and so on, to visible bike sheds.

Groups of the Design

Each design group consists of two designers.

Softwares of the Design

RhinoScript/Grasshopper

初步概念
Basic Concept

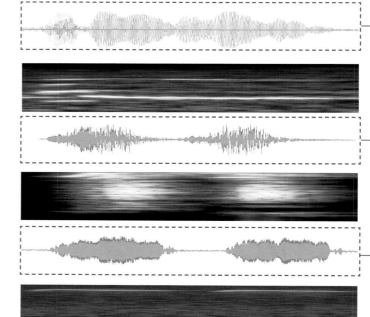

Architecture Program of SAUP, NJU 2011—2012

Graduate Program 1st Year
DIGITAL ARCHITECTURE DESIGN • JI Guohua

Type: Optional Course
Study Period and Credits: 36 hours/2 credits

形态的计算机编程
Computer Morphogenisis

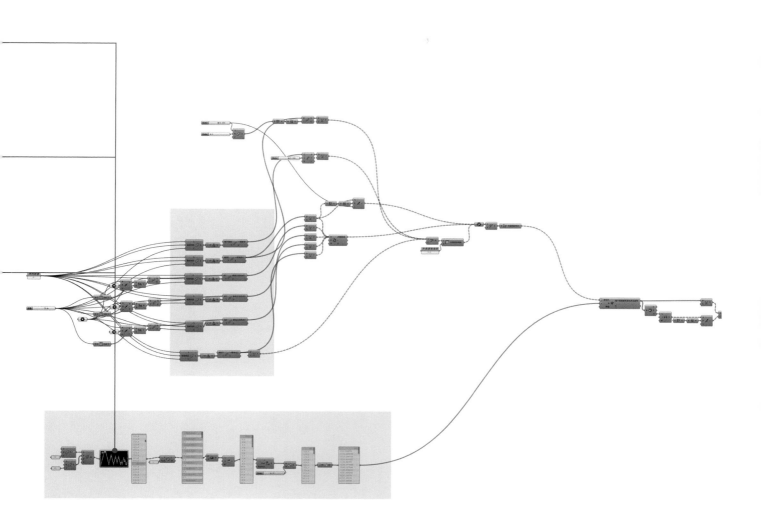

研究生一年级

建筑设计与构造设计
Architecture Design and Construction Design

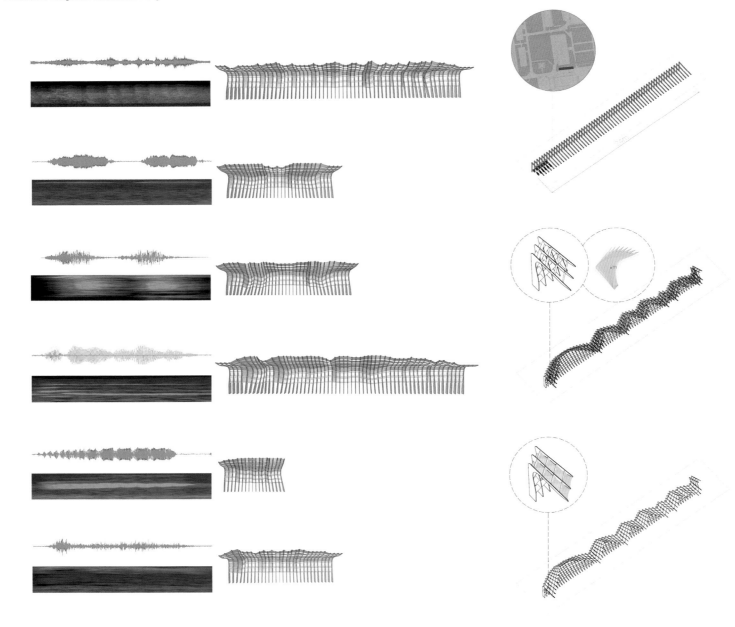

设计表达
Design Presentation

研究生一年级
影像南京 • Marc Boumeester

课程类型：选修
学时学分：18学时/1学分

Abstract:
The way we perceive our environment (or to be more precise, our habitat, being both physical and virtual) has changed very much in the past decades. We have only just begun to recognize the potential modern technologies (will) offer in the collaboration between people and things, especially when looking at their capacities (what they allow), rather than to emphasize their properties (what they are). Gilles Deleuze claims that cinema brings us movement-image; being the notion that bodies are not described in motion but the continuity of motion describes the object, "capable of thinking the production of the new". The workshop consists roughly of two parts; first we will try to map the moment-spaces in which we find affordances that allow us to interact with the city. On basis of these maps we will start to design speculative micro-interventions in this habitat, utilizing various montage techniques to re-enter the domain of the affect. Engaging in an investigation of the perceptual composition and constructed layering of classic cinema, combined with the hands-on production of contemporary audio-visuals, it aims to increase the understanding of the potential of this medium and its place in the contemporary discourse on immersive media.

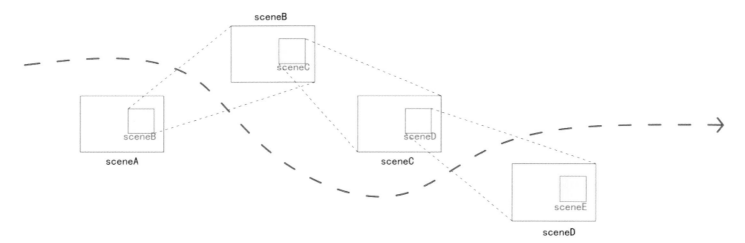

Requirements
Basic experience in operating a photo-digital-video camera, some knowledge/skills of editing (-software), mid-level understanding of English (B2-C1)

Reading
Graafland, Trafo (1991)
Pallasmaa, Lived Space in Architecture and Cinema (2007)

Optional reading
Guattari, The Three Ecologies (1998)
Ranciere, Dialectical Montage, Symbolic Montage (2007)

Architecture Program of SAUP, NJU 2011—2012

Graduate Program 1st Year
NANJING CINEMATIC • Marc Boumeester

Type: Optional Course
Study Period and Credits: 18 hours/1 credits

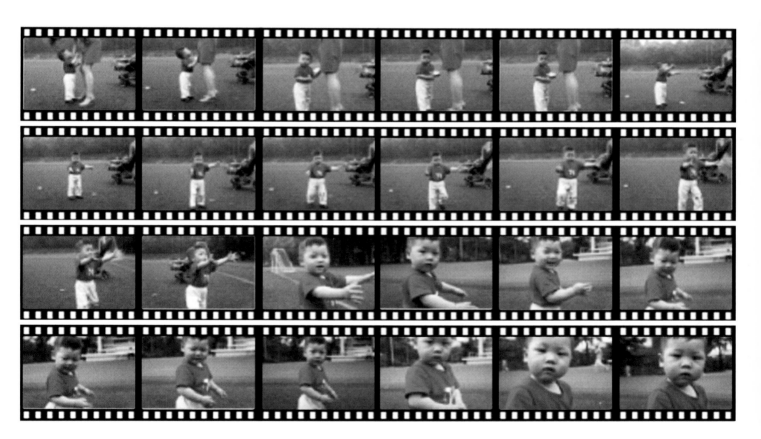

研究生一年级
体积编码（南京）• Tom Kvan & Justyna Karakiewicz

课程类型：选修
学时学分：18学时/1学分

基地
SITE

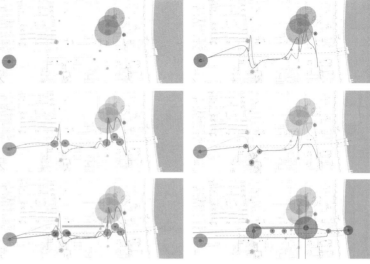

现状分析
PRESENT CONDITION ANALYSIS

In the first stage on the studio students will be asked to find, described, and demonstrate one specific site, located in close proximity to the University Campus, that in their mind illustrate the best the idea of under utilized space, inhospitable for pedestrian, but in the same time suggesting great opportunities for intervention, which may lead to dramatic change in the way this particular part of the city is used and experience.Possible sites maybe: traffic junctions, roundabouts; triangulated left over spaces resulted from two different grid structures intersecting with each other; car parks; commercial piazzas that do not perform as public spaces; spaces under flyover; any other under utilized spaces.We will select three sites for the further development.

The Second part of the studio will be related to defining and creating what we call program generator for the specific site. The students will be asked to use Grasshopper, or any other parametric software in order to demonstrate how their specific program could change the way in which the part of the city function. The students will be asked to allocated within their program to functions: a place for affordable living and a place for interaction and community. The rest of the mix will be left to interpretation of the individual student.

The final part of the studio will be related to creation of parametric diagram for the form, within urban structure.

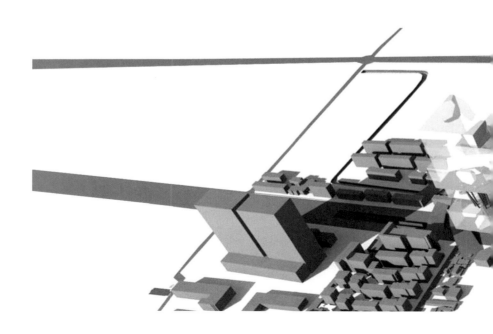

Graduate Program 1st Year
CODING VOLUMETRIC NANJING • Tom Kvan & Justyna Karakiewicz

Type: Optional Course
Study Period and Credits: 18 hours/1 credits

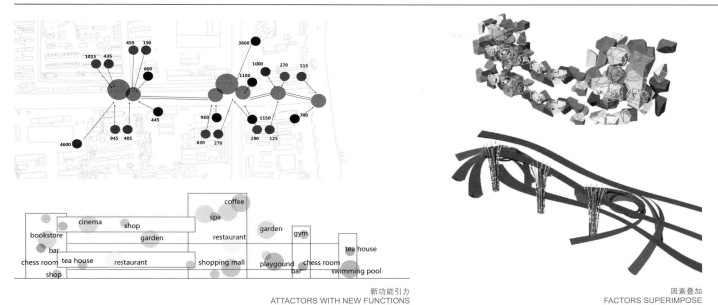

新功能引力
ATTACTORS WITH NEW FUNCTIONS

因素叠加
FACTORS SUPERIMPOSE

从"河"到"湖"
FROM RIVER TO LAKE

研究生一年级

虚拟餐厅 • Doris Fach & Hans Sebastian von Bernuth

课程类型：选修
学时学分：18学时/1学分

Program

In the country of the superlative, our studio takes interest in the small scale. We will dwell on the country's culture of eating and develop a descriptive design for a place to eat out. As the social performance of a festive menu or a fast food counter is highly dependent on the mood and feel created by the setting of it, we will focus on the creation of a distinct environment that is locally specific and consistent in all its elements. Dimensions and configurations of space, layout and design of furniture, color and material schemes, combinations of smooth and textured surfaces, lighting and signage, cutlery and tableware, will all be carefully arranged to achieve a recognizable atmosphere.

Site

Each group's site will be defined by one photograph of an urban situation in Nanjing. We will analyze this site by highlighting characteristic elements
and finally identify one building or structure that becomes home to the dining hall. Regardless of the actual interiors of that building, we will go on to imagine the dining hall behind its facade through a collage of images (mood board) that captures intuitive responses to the chosen site and complementary spatial and atmospheric associations. As a product of both the urban context and the associations it triggers, the environment of the dining hall exclusively depends on the character of the site and the team's imaginative creativity. The dining hall will take shape in one single space. This interior space will have certain relations with the pattern of apertures of its host building, but otherwise remain independent in scale and layout. In turn, the interior will make itself visible from the outside. Windows, doors, and immediate surroundings will be animated by the new function and eventually have an effect on the exterior façade of the host building and its larger urban context.

Deliverables

There are three main deliverables: one exterior image, one interior image, and one collection of close-ups. All three adhere to the format A0. Our work will thus evolve and develop through images only, leading us to focus on atmospheric and social qualities of space, instead of its Cartesian dimensions and logistics. Accordingly, we will produce neither plans, nor sections, details, or site-plans, but use all techniques at hand to create a holistic architectural impression. The interior image will, as defined above, start out as a collage of associations and imaginations and develop into a perspective, multi layered, and high-resolution presentation image during the course of the week. The exterior image will be based on the site photograph and experience adaptations of entry, apertures, signage, terrace, parking, etc. The collection of close-ups will highlight specific elements, materials, or thresholds of the proposed design at intimate focus.

最终室内图片
FINAL INTERIOR IMAGE

最终室外图片
FINAL EXTERIOR IMAGE

原始照片
ORIGINAL PHOTO

Graduate Program 1st Year

"EAT OUT": IMAGINATIONS OF A DINING HALL IN NANJING • Doris Fach & Hans Sebastian von Bernuth

Type: Optional Course
Study Period and Credits: 18 hours/1 credits

本科二年级
建筑导论・赵辰等
课程类型：必修
学时学分：36学时/2学分

Undergraduate Program 2nd Year
INTRODUCTORY GUIDE TO ARCHITECTURE
• ZHAO Chen, etc.
Type: Required Course
Study Period and Credits: 36 hours / 2 credits

课程内容
1. 建筑学的基本定义
　　第一讲：建筑与设计/赵辰
　　第二讲：建筑与城市/丁沃沃
　　第三讲：建筑与生活/张雷
2. 建筑的基本构成
2-1 建筑的物质构成
　　第四讲：建筑的物质环境/赵辰
　　第五讲：建筑与节能技术/秦孟昊
　　第六讲：建筑与人居环境/刘铨
　　第七讲：建筑与生态环境/吴蔚
　　第八讲：建筑与建造技术/冯金龙
2-2 建筑的文化构成
　　第九讲：建筑与人文、艺术、审美/赵辰
　　第十讲：建筑与环境景观/华晓宁
　　第十一讲：建筑与历史理论（西方）/胡恒
　　第十二讲：建筑与历史理论（中国古代）/萧红颜
　　第十三讲：建筑与历史理论（中国近代）/冷天
3. 建筑师职业与建筑学术
　　第十四讲：建筑与表现/赵辰
　　第十五讲：建筑与几何形态/周凌
　　第十六讲：建筑与数字技术/吉国华
　　第十七讲：城市与数字技术/童滋雨
　　第十八讲：建筑师的职业技能与社会责任/傅筱

Course Content
I Preliminary of architecture
1. Architecture and design / ZHAO Chen
2. Architecture and urbanization / DING Wowo
3. Architecture and life / ZHANG Lei
II Basic attribute of architecture
II-1 Physical attribute
4. Physical environment of architecture / ZHAO Chen
5. Architecture and energy saving / QIN Menghao
6. Architecture and habitation / LIU Quan
7. Architecture and ecological environment / WU Wei
8. Architecture and construction technology / FENG Jinlong
II-2 Cultural attribute
9. Architecture and civilization, arts, aesthetic / ZHAO Chen
10. Architecture and landscaping environment / HUA Xiaoning
11. The history and theory of architecture (Western) / HU Heng
12. The history and theory of architecture (Chinese Ancient) / XIAO Hongyan
13. The history and theory of architecture (Chinese Modern) / LENG Tian
III Architect: profession and academy
14. Architecture and presentation / ZHAO Chen
15. Architecture and geometrical form / ZHOU Ling
16. Architectural and digital technology / JI Guohua
17. Urban and digital technology / TONG Ziyu
18. Architect's professional technique and responsibility / FU Xiao

---

本科三年级
建筑设计基础原理・周凌
课程类型：必修
学时学分：36学时/2学分

Undergraduate Program 3rd Year
BASIC THEORY OF ARCHITECTURAL DESIGN
• ZHOU Ling
Type: Required Course
Study Period and Credits: 36hours / 2 credits

课程目标
　　本课程是建筑学专业本科生的专业基础理论课程。本课程的任务主要是介绍建筑设计中形式与类型的基本原理。形式原理包含历史上各个时期的设计原则，类型原理讨论不同类型建筑的设计原理。
课程内容
　　1. 形式与类型概述
　　2. 古典建筑形式语言
　　3. 现代建筑形式语言
　　4. 当代建筑形式语言
　　5. 类型设计
　　6. 材料与建造
　　7. 技术与规范
　　8. 课程总结
课程要求
　　1. 讲授大纲的重点内容；
　　2. 通过分析实例启迪学生的思维，加深学生对有关理论及其应用、工程实例等内容的理解；
　　3. 通过对实例的讨论，引导学生运用所学的专业理论知识，分析、解决实际问题。

Course Objective
This course is a basic theory course for the undergraduate students of architecture. The main purpose of this course is to introduce the basic principles of the form and type in architectural design. Form theory contains design principles in various periods of history; type theory discusses the design principles of different types of building.
Course Content
1 Overview of forms and types
2. Classical architecture form language
3. Modern architecture form language
4. Contemporary architecture form language
5. Type design
6. Materials and construction
7. Technology and specification
8. Course Summary
Course Requirement
1. Teach the key elements of the outline.
2. Enlighten students' thinking and enhance students' understanding of the theories, its applications and project examples through analyzing examples.
3. Guide students using the professional knowledge to analysis and solve practical problems through the discussion of examples.

本科三年级
居住建筑设计与居住区规划原理・冷天　刘铨
课程类型：必修
学时学分：36学时/2学分

Undergraduate Program 3rd Year
THEORY OF HOUSING DESIGN AND RESIDENYTIAL PLANNING • LENG Tian, LIU Quan
Type: Required Course
Study Period and Credits: 36 hours / 2 credits

课程内容
　　第一讲：课程概述
　　第二讲：居住建筑的演变
　　第三讲：套型空间的设计
　　第四讲：套型空间的组合与单体设计（一）
　　第五讲：套型空间的组合与单体设计（二）
　　第六讲：居住建筑的结构、设备与施工
　　第七讲：专题讲座：住宅的适应性、支撑体住宅
　　第八讲：城市规划理论概述
　　第九讲：现代居住区规划的发展历程
　　第十讲：居住区的空间组织
　　第十一讲：居住区的道路交通系统规划与设计
　　第十二讲：居住区的绿地景观系统规划与设计
　　第十三讲：居住区公共设施规划、竖向设计与管线综合
　　第十四讲：专题讲座：住宅产品开发
　　第十五讲：专题讲座：住宅产品设计实践
　　第十六讲：课程总结，考试答疑

Course Content
Lect. 1: Introduction of the course
Lect. 2: Development of residential building
Lect. 3: Design of dwelling space
Lect. 4: Dwelling space arrangement and residential building design (1)
Lect. 5: Dwelling space arrangement and residential building design (2)
Lect. 6: Structure, detail, facility and construction of residential buildings
Lect. 7: Adapt ability of residential building, supporting house
Lect. 8: Introduction of the theories of urban planning
Lect. 9: History of modern residential planning
Lect. 10: Organization of residential Space
Lect. 11: Traffic system planning and design of residential area
Lect. 12: Landscape planning and design of residential area
Lect. 13: Public facilities and infrastructure system
Lect. 14: Real estate development
Lect. 15: The practice of residential planning and housing design
Lect. 16: Summary, question of the test

研究生一年级
现代建筑设计基础理论·张雷
课程类型：必修
学时学分：18学时/1学分

Graduate Program 1st Year
PRELIMINARIES IN MODERN ARCHITECTURAL DESIGN • ZHANG Lei
Type: Required Course
Study Period and Credits:18 hours/1 credit

课程内容
1. 现代设计思想的演变
2. 基本空间的组织
3. 建筑类型的抽象和还原
4. 材料运用与建造问题
5. 场所的形成及其意义
6. 今天的工作原则与策略

建筑可以被抽象到最基本的空间围合状态来面对它所必须解决的基本的适用问题，用最合理、最直接的空间组织和建造方式去解决问题，以普通材料和通用方法去回应复杂的使用要求，是建筑设计所应该关注的基本原则。

Course Content
1. Transition of the modern thoughts of design
2. Arrangement of basic space
3. Abstraction and reversion of architectural types
4. Material application and constructional issues
5. Formation and significance of sites
6. Nowaday working principles and strategies

Architecture can be abstracted to the most fundamental state of space enclosure, so as to confront all the basic applicable problems which must be resolved. The most reasonable and direct mode of space arrangement and construction shall be applied; ordinary materials and universal methods shall be used as the countermeasures to the complicated application requirement. These are the basic principles on which an architecture design institution shall focus.

研究生一年级
现代建筑设计方法论·丁沃沃
课程类型：必修
学时学分：18学时/1学分

Graduate Program 1st Year
METHODOLOGY OF MODERN ARCHITECTURAL DESIGN • DING Wowo
Type: Required Course
Study Period and Credits:18 hours/1 credit

课程内容
以建筑历史为主线，讨论建筑设计方法演变的动因/理念及其方法论。基于对传统中国建筑和西方古典建筑观念异同的分析，探索方法方面的差异。通过分析建筑形式语言的逻辑关系，讨论建筑形式语言的几何学意义。最后，基于城市形态和城市空间的语境探讨了建筑学自治的意义。

1. 引言
2. 西方建筑学的传统
3. 中国:建筑的意义
4. 历史观与现代性
5. 现代建筑与意识的困境
6. 建筑形式语言的探索
7. 反思与回归理性
8. 结语

Course Content
Along the main line of architectural history,this course discussed the evolution of architectural design motivation ideas and methodology.Due to different concepts between the Chinese architecture and Western architecture Matters. The way for analyzing and exploring has to be studied.By analyzing the logic relationship of architectural form language,the geometrical significance of architectural form language is explored. Finally,within the context of urban form and space,the significance of architectural autonomy has been discussed.

1. Introduction
2. Tradition of western architecture
3. Meaning of architecture in china
4. History and modernity
5. Modern architectural ideology and its dilemma
6. Exploration for architectural form language
7. Re-thinking and return to reason
8. Conclusion

城市理论课程
URBAN THEORY

本科四年级
城市设计及其理论・丁沃沃 胡友培
课程类型：必修
学时学分：36学时/2学分

Undergraduate Program 4nd Year
THEORY OF URBAN DESIGN • DING Wowo, HU Youpei
Type: Required Course
Study Period and Credits: 36 hours / 2 credits

课程内容
第一讲 课程概述
第二讲 城市设计技术术语：城市规划相关术语；城市形态相关术语；城市交通相关术语；消防相关术语
第三讲 城市设计方法 —— 文本分析：城市设计上位规划；城市设计相关文献；文献分析方法
第四讲 城市设计方法 —— 数据分析：人口数据分析与配置；交通流量数据分析；功能分配数据分析；视线与高度数据分析；城市空间数据模型的建构
第五讲 城市设计方法 —— 城市肌理分类：城市肌理分类概述；肌理形态与建筑容量；肌理形态与开放空间；肌理形态与交通流量；城市绿地指标体系
第六讲 城市设计方法 —— 城市路网组织：城市道路结构与交通结构概述；城市路网与城市功能；城市路网与城市空间；城市路网与市政设施；城市道路断面设计
第七讲 城市设计方法 —— 城市设计表现：城市设计分析图；城市设计概念表达；城市设计成果解析图；城市设计地块深化设计表达；城市设计空间表达
第八讲 城市设计的历史与理论：城市设计的历史意义；城市设计理论的内涵
第九讲 城市路网形态：路网形态的类型和结构；路网形态与肌理；路网形态的变迁
第十讲 城市空间：城市空间的类型；城市空间结构；城市空间形态；城市空间形态的变迁
第十一讲 城市形态学：英国学派；意大利学派；法国学派；空间句法
第十二讲 城市形态的物理环境：城市形态与物理环境；城市形态与环境研究；城市形态与环境测评；城市形态与环境操作
第十三讲 景观都市主义：景观都市主义的理论、操作和范例
第十四讲 城市自组织现象及其研究：城市自组织现象的魅力与问题；城市自组织系统研究方法；典型自组织现象案例研究
第十五讲 建筑学图式理论与方法：图式理论的研究，建筑学图式的概念；图式理论的应用；作为设计工具的图式；当代城市语境中的建筑学图式理论探索
第十六讲 课程总结

Course content
Lect. 1. Introduction
Lect. 2. Technical terms: terms of urban planning, urban morphology, urban traffic and fire protection.
Lect. 3. Urban design methods — documents analysis: urban planning and policies; relative documents; document analysis techniques and skills.
Lect. 4. Urban design methods — data analysis: data analysis of demography, traffic flow, public facilities distribution, visual and building height; modelling urban spatial data.
Lect. 5. Urban design methods — classification of urban fabrics: introduction of urban fabrics; urban fabrics and floor area ratio; urban fabrics and open space; urban fabrics and traffic flow; criteria system of urban green space.
Lect. 6. Urban design methods — organization of urban road network: introduction; urban road network and urban function; urban road network and urban space; urban road network and civic facilities; design of urban road section.
Lect. 7. Urban design methods — representation skills of urban Design: mapping and analysis; conceptual diagram; analytical representation of urban design; representation of detail design; spatial representation of urban design.
Lect. 8. Brief history and theories of urban design: historical meaning of urban design; connotation of urban design theories.
Lect. 9. Form of urban road network: typology, structure and evolution of road network; road network and urban fabrics.
Lect. 10. Urban space: typology, structure, morphology and evolution of urban space.
Lect. 11. Urban morphology: Cozen School; Italian School; French School; Space Syntax theory.
Lect. 12. Physical environment of urban forms: urban forms and physical environment; environmental study; environmental evaluation and environmental operations.
Lect. 13. Landscape urbanism: ideas, theories, operations and examples of landscape urbanism.
Lect. 14. Researches on the phenomena of the urban self-Organization: charms and problems of urban self-organization phenomena; research methodology on urban self-organization phenomena; case studies of urban self-organization phenomena.
Lect. 15. Theory and method of architectural diagram: theoretical study on diagrams; concepts of architectural diagrams; application of diagram theory; diagrams as design tools; theoretical research of architectural diagrams in contemporary urban context.
Lect. 16. Summary

研究生一年级
城市形态研究・丁沃沃 赵辰 萧红颜
课程类型：必修
学时学分：36学时/2学分

Graduate Program 1nd Year
URBAN MORPHOLOGY • DING Wowo, ZHAO Chen, XIAO Hongyan
Type: Required Course
Study Period and Credits: 36 hours / 2 credits

课程要求
1.要求学生基于对历史性城市形态的认知分析，加深对中西方城市理论与历史的理解。
2.要求学生基于历史性城市地段的形态分析，提高对中西方城市空间特质及相关理论的认知能力。

课程内容
第一周：序言 概念、方法及成果
第二周：讲座1 城市形态认知的历史基础 —— 营造观念与技术传承
第三周：讲座2 城市形态认知的历史基础 —— 图文并置与意象构建
第四周：讲座3 城市形态认知的理论基础 —— 价值判断与空间生产
第五周：讲座4 城市形态认知的理论基础 —— 勾沉呈现与特征形塑
第六周：讲座5 历史城市的肌理研究
第七周：讲座6 整体与局部 —— 建筑与城市
第八周：讨论
第九周：讲座7 城市化与城市形态
第十周：讲座8 城市乌托邦
第十一周：讲座9 走出乌托邦
第十二周：讲座10 重新认识城市
第十三周：讲座11 城市设计背景
第十四周：讲座12 城市设计实践
第十五周：讲座13 城市设计理论
第十六周：讲评

Course Requirement
1. Deepen the understanding of Chinese and Western urban theories and histories based on the cognition and analysis of historical urban form.
2. Improve the cognitive abilities of the characteristics and theories of Chinese and Western urban space based on the morphological analysis of the historical urban sites.
Course Content
Week 1. Preface — concepts, methods and results
Week 2. Lect. 1 Historical basis of urban form cognition — Developing concepts and passing of technologies
Week 3. Lect. 2 Historical basis of urban form cognition — Apposition of pictures and text and imago construction
Week 4. Lect. 3 Theoretical basis of urban form cognition — Value judgement and space production
Week 5. Lect. 4 Theoretical basis of urban form cognition — History representation and feature shaping
Week 6. Lect. 5 Study on the grain of historical cities
Week 7. Lect. 6 Whole and part: Architecture and urban
Week 8. Discussion
Week 9. Lect. 7 Urbanization and urban form
Week 10. Lect. 8 Urban utopia
Week 11. Lect. 9 Walk out of Utopia
Week 12. Lect. 10 Have a new look of the city
Week 13. Lect. 11 Background of urban design
Week 14. Lect. 12 Practice of urban design
Week 15. Lect. 13 Theory of urban design
Week 16. Discussions

本科四年级
景观规划设计及其理论·尹航
课程类型：选修
学时学分：36学时/2学分

Undergraduate Program 4th Year
LANDSCAPE PALNNING DESIGN AND THEORY
• YIN Hang
Type: Elective Course
Study Period and Credits: 36 hours / 2 credits

课程介绍
　　景观规划设计的对象包括所有的室外环境，景观与建筑的关系往往是紧密而互相影响的，这种关系在城市中尤为明显。景观规划设计及理论课程希望从景观设计理念、场地设计技术、建筑周边环境塑造等方面开展课程的教学，为建筑学本科生建立更加全面的景观知识体系，并且完善建筑学本科生在建筑场地设计、总平面规划与城市设计等方面的设计能力。
　　本课程主要从三个方面展开：一是理念与历史：以历史的视角介绍景观学科的发展过程，让学生对景观学科有一个宏观的了解，初步理解景观设计理念的发展；二是场地与文脉：通过阐述景观规划设计与周边自然环境、地理位置、历史文脉和方案可持续性的关系，建立场地与文脉的设计思维；三是景观与建筑：通过设计方法授课、先例分析作业等方式让学生增强建筑的环境意识，了解建筑的场地设计的影响因素、一般步骤与设计方法，并通过与"建筑设计6"和"建筑设计7"的设计任务书相配合的同步课程设计训练来加强学生景观规划设计的能力。

Course Description
The object of landscape planning design includes all outdoor environments; the relationship between landscape and building is often close and interactive, which is especially obvious in a city. This course expects to carry out teaching from perspective of landscape design concept, site design technology, building's peripheral environment creation, etc. to establish a more comprehensive landscape knowledge system for the undergraduate students of architecture, and perfect their design ability in building site design, master plane planning and urban design and so on.
This course includes three aspects:
1. Concept and history;
2. Site and context;
3. Landscape and building.

本科四年级
东西方园林·许浩
课程类型：选修
学时学分：36学时/2学分

Undergraduate Program 4th Year
GARDEN OF EAST AND WEST • XU Hao
Type: Elective Course
Study Period and Credits: 36 hours / 2 credits

课程介绍
　　帮助学生系统掌握园林、绿地的基本概念、理论和研究方法，尤其了解园林艺术的发展脉络，侧重各个流派如日式园林、江南私家园林、皇家园林、规则式园林、自由式园林、伊斯兰园林的不同特征和关系；使得学生能够从社会背景、环境等方面解读园林的发展特征，并能够开展一定的评价。

Course Description
Help students systematically master the basic concepts, theories and research methods of gardens and greenbelts, especially understand the evolution of gardening, emphasizing the different features and relationships of various genres, such as Japanese gardens, private gardens by the south of Yangtze River, royal gardens, rule-style gardens, free style gardens and Islamic gardens; enable students to interpret the characteristics of garden development from the point of view of social backgrounds, environment, etc. Furthermore to do the evaluation in depth.

研究生一年级
景观规划进展·许浩
课程类型：选修
学时学分：18学时/1学分

Graduate Program 1st Year
LANDSCAPE PLANNING PROGRESS • XU Hao
Type: Elective Course
Study Period and Credits: 18 hours / 1 credits

课程介绍
　　生态规划是景观规划的核心内容之一。本课程总结了生态系统、生态保护和生态修复的基本概念。大规模生态保护的基本途径是国家公园体系，而生态修复则是通过人为干涉对破损环境的恢复。本课程介绍了国家公园的价值、分类与成就，并通过具体案例论述了欧洲、澳洲景观设计过程中生态修复的做法。

Course Description
Ecological planning is one of the core contents of landscape planning. This course summarizes the basic concepts of ecological systems, ecological protection and ecological restoration. The basic channel of large-scale ecological protection is the national park system, while ecological restoration is to restore the damaged environment by means of human intervention. This course introduces the values, classification and achievements of national parks, and discusses the practices of ecological restoration in the process of landscape design in Europe and Australia.

研究生一年级
景观都市主义理论与方法·华晓宁
课程类型：选修
学时学分：18学时/1学分

Graduate Program 1st Year
THEORY AND METHOD OF LANDSCAPE URBANISM
• HUA Xiaoning
Type: Elective Course
Study Period and Credits: 18 hours / 1 credits

课程介绍
　　本课程介绍了景观都市主义思想产生的背景、缘起及其主要理论观点，并结合实例，重点分析了其在不同的场址和任务导向下发展起来的多样化的实践策略和操作性工具。通过这些内容的讲授，本课程的最终目的是拓宽学生的视野，引导学生改变既往的思维定式，以新的学科交叉整合的思路，分析和解决当代城市问题。

课程内容
　　第一讲：导论——当代城市与景观媒介
　　第二讲：生态过程与景观修复
　　第三讲：基础设施与景观嫁接
　　第四讲：嵌入与缝合
　　第五讲：水平性与都市表面
　　第六讲：城市图绘与图解
　　第七讲：AA景观都市主义——原型方法
　　第八讲：总结与作业

Course Description
The course introduces the backgrounds, the generation and the main theoretical opinions of landscape urbanism. With a series of instances, it particularly analyses the various practical strategies and operational techniques guided by various sites and projects. With all these contents, the aim of the course is to widen the students' field of vision, change their habitual thinking and suggest them to analyze and solve contemporary urban problems using the new ideas of the intersection and integration of different disciplines.

Course Content
Lect. 1: Introduction — contemporary cities and landscape medium
Lect. 2: Ecological process and landscape recovering
Lect. 3: Infrastructure and landscape engrafting
Lect. 4: Embedment and oversewing
Lect. 5: Horizontality and urban surface
Lect. 6: Urban mapping and diagram
Lect. 7: AA Landscape Urbanism — archetypical method
Lect. 8: Conclusion and assignment

历史理论课程
HISTORY THEORY

本科二年级
中国建筑史（古代）・萧红颜
课程类型：必修
学时学分：36学时/2学分

Undergraduate Program 2nd Year
HISTORY OF CHINESE ARCHITECTURE (ANCIENT)
• XIAO Hongyan
Type: Required Course
Study Period and Credits: 36 hours / 2 credits

课程目标
认识中国传统营造的思维特征与技术选择，培养学生理解历史与分析问题的意识。

课程内容
采取有别于传统教学的方式，将专题讲述与类型分析并重，强调多元视角下的问题式教学，力求达成认知与学理的综合效应。

Course Objective
Recognize the thinking characteristics and technology selection of China's traditional construction; develop students' awareness of understanding history and analyzing problems.
Course Content
Use a teaching method different from traditional teaching, concurrently emphasizes special lecturing and type analysis as well as the problem-based teaching under diversified visions, attempting to realize the integrated effect of cognition and learnt theory.

本科二年级
外国建筑史（古代）・胡恒
课程类型：必修
学时学分：36学时/2学分

Undergraduate Program 2nd Year
HISTORY OF WESTERN ARCHITECTURE (ANCIENT) •
HU Heng
Type: Required Course
Study Period and Credits: 36 hours / 2 credits

课程目标
本课程力图对西方建筑史的脉络做一整体勾勒，使学生在掌握重要的建筑史知识点的同时，对西方建筑史在2000多年里的变迁的结构转折（不同风格的演变）有所深入的理解。本课程希望学生对建筑史的发展与人类文明发展之间的密切关联有所认识。

课程内容
1. 概论 2. 希腊建筑 3. 罗马建筑 4. 中世纪建筑
5. 意大利的中世纪建筑 6. 文艺复兴 7. 巴洛克
8. 美国城市 9. 北欧浪漫主义 10. 加泰罗尼亚建筑
11. 先锋派 12. 德意志制造联盟与包豪斯
13. 苏维埃的建筑与城市 14. 20世纪60年代的建筑
15. 20世纪70年代的建筑 16. 答疑

Course Objective
This course seeks to give an overall outline of Western architectural history, so that the students may have an in-depth understanding of the structural transition (different styles of evolution) of Western architectural history in the past 2000 years. This course hopes that students can understand the close association between the development of architectural history and the development of human civilization.
Course Content
1. Generality 2. Greek Architectures 3. Roman Architectures
4. The Middle Ages Architectures
5. The Middle Ages Architectures in Italy 6. Renaissance
7. Baroque 8. American Cities 9. Nordic Romanticism
10. Catalonian Architectures 11. Avant-Garde
12. German Manufacturing Alliance and Bauhaus
13. Soviet Architecture and Cities 14. 1960's Architectures
15. 1970's Architectures 16. Answer Questions

本科三年级
外国建筑史（当代）・胡恒
课程类型：必修
学时学分：36学时/2学分

Undergraduate Program 3rd Year
HISTORY OF WESTERN ARCHITECTURE (MODERN)
• HU Heng
Type: Required Course
Study Period and Credits: 36 hours / 2 credits

课程目标
本课程力图用专题的方式对文艺复兴时期的7位代表性的建筑师与5位现当代的重要建筑师作品做一细致的讲解。本课程将重要建筑师的全部作品尽可能在课程中梳理一遍，使学生能够全面掌握重要建筑师的设计思想、理论主旨、与时代的特殊关联、在建筑史中的意义。

课程内容
1. 伯鲁乃涅斯基 2. 阿尔伯蒂 3. 伯拉孟特
4. 米开朗琪罗（1） 5. 米开朗琪罗（2） 6. 罗马诺
7. 桑索维诺 8. 帕拉蒂奥（1） 9. 帕拉蒂奥（2）
10. 赖特 11. 密斯 12. 勒·柯布西耶（1）
13. 勒·柯布西耶（2） 14. 海杜克 15. 妹岛和世
16. 答疑

Course Objective
This course seeks to make a detailed explanation to the works of 7 representative architects in the Renaissance period and 5 important contemporary architects in a special way. This course will try to reorganize all works of these important architects, so that the students can fully grasp their design ideas, theoretical subject and their particular relevance with the era and significance in the architectural history.
Course Content
1. Brunelleschi 2. Alberti 3. Bramante
4. Michelangelo(1) 5. Michelangelo(2)
6. Romano 7. Sansovino 8. Paratio(1) 9. Paratio(2)
10. Wright 11. Mies 12. Le Corbusier(1) 13. Le Corbusier(2)
14. Hejduk 15. Kazuyo Sejima
16. Answer Questions

本科三年级
中国建筑史（近现代）・赵辰
课程类型：必修
学时学分：36学时/2学分

Undergraduate Program 3rd Year
HISTORY OF CHINESE ARCHITECTURE (MODERN)
• ZHAO Chen
Type: Required Course
Study Period and Credits: 36 hours / 2 credits

课程介绍
本课程作为本科建筑学专业的历史与理论课程，是中国建筑史教学中的一部分。在中国与西方的古代建筑历史课程先学的基础上，了解中国社会进入近代，以至于现当代的发展进程。
在对比中西方建筑文化的基础之上，建立对中国近现代建筑的整体认识。深刻理解中国传统建筑文化在近代以来与西方建筑文化的冲突与相融之下，逐步演变发展至今天世界建筑文化之一部分之意义。

Course Description
As the history and theory course for undergraduate students of Architecture, this course is part of the teaching of History of Chinese Architecture. Based on the earlier studying of Chinese and Western history of ancient architecture, understand the evolution progress of Chinese society's entry into modern times and even contemporary age.
Based on the comparison of Chinese and Western building culture, establish the overall understanding of China's modern and contemporary buildings. Have further understanding of the significance of China's traditional building culture's gradual evolution into one part of today's world building culture under conflict and blending with Western building culture in modern times.

研究生一年级
建筑理论研究 · 王骏阳
课程类型：必修
学时学分：18学时/1学分

Graduate Program 1st Year
STUDY OF ARCHITECTURAL THEORY · WANG Junyang
Type: Required Course
Study Period and Credits:18 hours / 1 credits

课程介绍
　　本课程是西方建筑史研究生教学的一部分。主要涉及当代西方建筑界具有代表性的思想和理论，其主题包括历史主义、先锋建筑、批判理论、建构文化以及当代城市的解读等等。本课程大量运用图片资料，广泛涉及哲学、历史、艺术等领域，力求在西方文化发展的背景中呈现建筑思想和理论的相对独立性及关联性，理解建筑作为一种人类活动所具有的社会和文化意义，启发学生的理论思维和批判精神。

课程内容
　　第一讲 建筑理论概论
　　第二讲 建筑自治
　　第三讲 柯林·罗：理想别墅的数学与其他
　　第四讲 阿道夫·路斯与装饰美学
　　第五讲 库哈斯与当代城市的解读
　　第六讲 意识的困境：对现代建筑的反思
　　第七讲 弗兰普顿的建构文化研究
　　第八讲 现象学

Course Description
This course is a part of teaching Western architectural history for graduate students. It mainly deals with the representative thoughts and theories in Western architectural circles, including historicism, vanguard building, critical theory, construction culture and interpretation of contemporary cities and more. Using a lot of pictures involving extensive fields including philosophy, history, art, etc., this course attempts to show the relative independence and relevance of architectural thoughts and theories under the development background of Western culture, understand the social and cultural significance owned by architectures as human activities, and inspire students' theoretical thinking and critical spirit.

Course Content
Lect. 1. Introduction to architectural theories
Lect. 2. Autonomous architecture
Lect. 3. Colin Rowe : the mathematics of the ideal villa and others
Lect. 4. Adolf Loos and adornment aesthetics
Lect. 5. Koolhaas and the interpretation of con-temporary cities
Lect. 6. Conscious dilemma: the reflection of modern architecture
Lect. 7. Studies in tectonic culture of Frampton
Lect. 8. Phenomenology

研究生一年级
建筑史研究 · 胡恒
课程类型：选修
学时学分：36学时/2学分

Graduate Program 1st Year
ARCHITECTURAL HISTORY RESEARCH · HU Heng
Type: Elective Course
Study Period and Credits:36 hours / 2 credits

课程目标
　　本课程的目的有二。其一，通过对建筑史研究的方法做一概述，来使学生粗略了解西方建筑史研究方法的总的状况。其二，通过对当代史概念的提出，且用若干具体的案例研究，来向学生展示当代史研究的路数、角度、概念定义、结构布置、主题设定等内容。

课程内容
　　1. 建筑史方法概述（1）
　　2. 建筑史方法概述（2）
　　3. 建筑史方法概述（3）
　　4. 塔夫里的建筑史研究方法
　　5. 当代史研究方法——周期
　　6. 当代史研究方法——杂交
　　7. 当代史研究方法——阈限
　　8. 当代史研究方法——对立

Course Objective
This course has two objectives: 1. Give the students a rough understanding of the overall status of the research approaches of the Western architectural history through an overview of them. 2. Show students the approaches, point of view, concept definition, structure layout, theme settings and so on of contemporary history study through proposing the concept of contemporary history and several case studies.

Course Content
1. The overview of the method of architectural history(1)
2. The overview of the method of architectural history(2)
3. The overview of the method of architectural history(3)
4. Tafuri's study method of architectural history
5. The study method of contemporary history — period
6. The study method of contemporary history — hybridization
7. The study method of contemporary history — limen
8. The study method of contemporary history — opposition

研究生一年级
建筑史研究 · 萧红颜
课程类型：选修
学时学分：36学时/2学分

Graduate Program 1st Year
ARCHITECTURAL HISTORY RESEARCH · XIAO Hongyan
Type: Elective Course
Study Period and Credits:36 hours / 2 credits

课程目标
　　本课程尝试从理念与类型两大范畴为切入点，专题讲述中国传统营造基本理念之嬗变与延续、基本类型之关联与意蕴，强调建筑史应回归艺术史分析框架下阐发相关史证问题及其方法。

课程内容
　　1. 边角　　2. 堪舆
　　3. 界域　　4. 传摹
　　5. 宫台　　6. 池苑
　　7. 庙墓　　8. 楼亭

Course Objective
This course attempts to start with two areas (concept and type) to state the mutation and continuation, association and implication of basic types of the basic concept of China's traditional construction, emphasizing that the architectural history should return to the art history framework to state relevant history evidence issues and its methods.

Course Content
1. Corner　　　　　　　　　2. Geomantic Omen
3. Boundary　　　　　　　　4. The Spreading and Copying
5. The Table Land of Palace　6. Pond
7. The Temple and Mausoleum　8. The Storied Building Pavilion

建筑技术课程
ARCHITECTURAL TECHNOLOGY

本科二年级
CAAD理论与实践・童滋雨
课程类型：必修
学时学分：36学时/2学分

Undergraduate Program 2nd Year
THEORY AND PRACTICE OF CAAD • TONG Ziyu
Type: Required Course
Study Period and Credits: 36 hours / 2 credits

课程介绍
在现阶段的CAD教学中，强调了建筑设计在建筑学教学中的主干地位，将计算机技术定位于绘图工具，本课程就是帮助学生可以尽快并且熟练地掌握如何利用计算机工具进行建筑设计的表达。课程中整合了CAD知识、建筑制图知识以及建筑表现知识，将传统CAD教学中教会学生用计算机绘图的模式向教会学生用计算机绘制有形式感的建筑图的模式转变，强调准确性和表现力作为评价CAD学习的两个最重要指标。
本课程的具体学习内容包括：
1. 初步掌握AutoCAD软件和SketchUP软件的使用，能够熟练完成二维制图和三维建模的操作；
2. 掌握建筑制图的相关知识，包括建筑投影的基本概念，平立剖面、轴测、透视和阴影的制图方法和技巧；
3. 图面效果表达的技巧，包括黑白线条图和彩色图纸的表达方法和排版方法。

Course Description
The core position of architectural design is emphasized in the CAD course. The computer technology is defined as drawing instrument. The course helps students learn how to make architectural presentation using computer fast and expertly. The knowledge of CAD, architectural drawing and architectural presentation are integrated into the course. The traditional mode of teaching students to draw in CAD course will be transformed into teaching students to draw architectural drawing with sense of form. The precision and expression will be emphasized as two most important factors to estimate the teaching effect of CAD course.
Contents of the course includes:
1. Use AutoCAD and SketchUP to achieve the 2-D drawing and 3-D modeling expertly.
2. Learn relational knowledge of architectural drawing, including basic concepts of architectural projection, drawing methods and skills of plan, elevation, section, axonometry, perspective and shadow.
3. Skills of presentation, including the methods of expression and lay out using mono and colorful drawings

---

本科三年级
建筑技术 1 — 结构、构造与施工・傅筱
课程类型：必修
学时学分：36学时/2学分

Undergraduate Program 3rd Year
ARCHITECTURAL TECHNOLOGY 1 - STRUCTURE, CONSTRUCTION AND EXECUTION • FU Xiao
Type: Required Course
Study Period and Credits:36 hours / 2 credits

教学目标
本课程是建筑学专业本科生的专业主干课程。本课程的任务主要是以建筑师的工作性质为基础，讨论一个建筑生成过程中最基本的三大技术支撑（结构、构造、施工）的原理性知识要点，以及它们在建筑实践中的相互关系。

Course Objective
The course is a major course for the undergraduate students of architecture. The main purpose of this course is based on the nature of the architect's work, to discuss the principle knowledge points of the basic three technical supports in the process of generating construction (structure, construction, execution), and their mutual relations in the architectural practice.

本科三年级
建筑技术 2 — 建筑物理・吴蔚
课程类型：必修
学时学分：36学时/2学分

Undergraduate Program 3rd Year
ARCHITECTURAL TECHNOLOGY 2 — BUILDING PHYSICS • WU Wei
Type: Required Course
Study Period and Credits:36 hours / 2 credits

课程介绍
本课程是针对三年级学生所设计，课程介绍了建筑热工学、建筑光学、建筑声学中的基本概念和基本原理，使学生能掌握建筑的热环境、声环境、光环境的基本评估方法，以及相关的国家标准。完成学业后在此方向上能阅读相关书籍，具备在数字技术方法等相关资料的帮助下，完成一定的建筑节能设计的能力。

课程目标
1. 了解建筑热工学、建筑光学、建筑声学中的基本概念和基本原理；
2. 掌握建筑的热环境、声环境、光环境的质量评价方法；
3. 了解和掌握计算机模拟技术在建筑节能方面的应用；
4. 了解建筑节能设计的基本原则和理论，并综合利用在设计中。

Course Description
Designed for the Grade-3 students, this course introduces the basic concepts and basic principles in architectural thermal engineering, architectural optics and architectural acoustics, so that the students can master the basic methods for the assessment of building's thermal environment, sound environment and light environment as well as the related national standards. After graduation, the students will be able to read the related books regarding these aspects, and have the ability to complete certain building energy efficiency designs with the help of the related digital techniques and methods.
Course Objective
1. Understand the basic concepts and principles in building thermal engineering, building optics and building acoustic;
2. Learn the quality assessment methods of building's thermal environment, sound environment and light environment;
3. Understand and learn the application of computer analysing technique in building energy efficiency;
4. Understand the basic principles and theories of building energy efficiency design, and use them in the design comprehensively.

本科三年级
建筑技术 3 — 建筑设备・吴蔚
课程类型：必修
学时学分：36学时/2学分

Undergraduate 3rd Year
ARCHITECTURAL TECHNOLOGY 3 — BUILDING EQUIPMENT • WU Wei
Type: Required Course
Study Period and Credits:36 hours / 2 credits

课程介绍
本课程是针对南京大学建筑与城市规划学院本科学生三年级所设计。课程介绍了建筑给水排水系统、采暖通风与空气调节系统、电气工程的基本理论、基本知识和基本技能，使学生能熟练地阅读水电、暖通工程图，熟悉水电及消防的设计、施工规范，了解燃气供应、安全用电及建筑防火、防雷的初步知识。

课程目标
1. 掌握各类建筑设备系统的特性、系统要求、系统布置，以及与其建筑物的相互关系；
2. 能够阅读建筑给排水、电气、暖通空调工程图，了解建筑师与设备工程师在实际工作中的合作关系；
3. 能够评估与解决与建筑设备相关的技术、经济、能源、环境影响的特定问题。

Course Description
This course is an undergraduate class offered in the School of Architecture and Urban Planning, Nanjing University. It introduces the basic principle of the building services systems, the technique of integration amongst the building services and the building. Throughout the course, the fundamental important to energy, ventilation, air-conditioning and comfort in buildings are highlighted.
Course Objective
1. A good understanding of the major building services systems and their integration and coordination into the architecture;
2. A good understanding of the working relationship between the architects and the building services engineers professor practice;
3. An ability to assess and solve some particular problems, which relate to technical performance, economics, energy usage and environmental effect.

研究生一年级
建筑节能与可持续发展・秦孟昊
课程类型：选修
学时学分：18学时/1学分

Graduate Program 1st Year
ENERGY CONSERVATION AND SUSTAINABLE ARCHITECTURE • QIN Menghao
Type: Elective Course
Study Period and Credits:18 hours / 1 credits

课程介绍
随着我国建筑总量的不断攀升和居住舒适度的提高，建筑能耗急剧上升。建筑节能已经为影响能源安全和提高能源利用效率的重要因素之一。建筑节能的关键首先是要设计"本身节能的建筑"，这样就能够从根本上减少建筑对高耗能建筑设备的依赖，从而减少建筑能耗。建筑师必须从建筑设计的最初阶段就进行规划，在建筑的形体、结构、开窗方式、外墙选材等方面融入节能设计的定量分析。而这些很难通过传统建筑设计方法达到，必须依靠建筑技术、建筑设备等多学科互动协作才能完成。但是目前很多建筑节能技术由设备工程师开发，由于专业背景不同，建筑师在设计过程中对这些技术了解不够，使用有一定难度。加强建筑设计专业和建筑技术专业在设计过程中的互动，成为世界各大建筑与城市规划学院本科教学的一个重点。
本课程将采用双语教学，主要面向建筑设计专业学生讲授建筑物理、建筑技术专业关于建筑节能方面的基本理念、设计方法和模拟软件，并指导学生将这些知识运用到节能建筑设计的过程中，在建筑设计专业和建筑技术专业之间建立一个互动的平台，从而达到设计"绿色建筑"的目标。教学过程中还将邀请外国著名专家为同学们介绍专业知识和互动设计的方法，培养学生参与国际合作互动设计的积极性，为以后开展交叉学科研究，培养复合型人才奠定基础。

Course Description
With the rising of China's total number of buildings and the need for living comfort, building energy consumption is rising sharply. Building energy efficiency has become one of the key factors influencing the energy security and energy efficiency. The first key for building energy efficiency is to design "a building that conserves energy itself", in this way, the reliance of building with high energy consuming equipment is essentially reduced, thus reducing the building's energy consumption. Architects must carry out planning at the very beginning of building design, and incorporate quantitative analysis of energy conservation into such aspects as the building's form, structure, window opening method and outer wall material selection and so on. However, it is difficult to satisfy them by means of traditional architectural design approaches; it must be realized by interactive collaboration of diversified subjects including construction technology, construction equipment, etc. However, most building energy efficiency technologies are developed by equipment engineers. Due to different professional backgrounds, architects have difficulty in using them in designing because of a lack of understanding of them. Strengthening the interaction of architectural design specialties and construction technology specialties in designing has become a key point in the undergraduate teaching in various large architecture and urban planning colleges around the world.

研究生一年级
材料与建造・冯金龙
课程类型：必修
学时学分：18学时/1学分

Graduate Program 1st Year
MATERIAL AND CONSTRUCTION • FENG Jinlong
Type: Required Course
Study Period and Credits:18 hours / 1 credits

课程介绍
介绍现代建筑技术的发展过程，论述现代建筑技术及其美学观念对建筑设计的重要作用。探讨由材料、结构和构造方式所形成的建筑建造的逻辑方式研究。研究建筑形式产生的物质技术基础，诠释现代建筑的建构理论与研究方法。

Course Description
It introduces the development process of modern architecture technology and discusses the important role played by the modern architecture technology and its aesthetic concepts in the architectural design. It explores the logical methods of construction of the architecture formed by materials, structure and construction. It studies the material and technical basis for the creation of architectural form, and interprets construction theory and research methods for modern architectures.

研究生一年级
计算机辅助技术・吉国华
课程类型：选修
学时学分：36学时/2学分

Graduate Program 1st Year
TECHNOLOGY OF CAAD • JI Guohua
Type: Elective Course
Study Period and Credits:36 hours / 2 credits

课程介绍
随着计算机辅助建筑设计技术的快速发展，当前数字技术在建筑设计中的角色逐渐从辅助绘图转向了真正的辅助设计，并引发了设计的革命和建筑的形式创新。本课程讲授AutoCAD VBA和RhinoScript编程。让学生在掌握"宏"/"脚本"编程的同时，增强以理性的过程思维方式分析和解决设计问题的能力，为数字建筑设计打下必要的基础。
课程分为三个部分：
1. VB语言基础，包括VB基本语法、结构化程序、数组、过程等编程知识和技巧；
2. AutoCAD VBA，包括AutoCAD VBA的结构、二维图形、人机交互、三维对象等，以及基本的图形学知识；
3. RhinoScript概要，包括基本概念、Nurbs概念、VBScript简介、曲线对象、曲面对象等。

Course Description
Following its fast development, the role of digital technology in architecture is changing from computer-aided drawing to real computer-aided design, leading to a revolution of design and the innovation of architectural form. Teaching the programming with AutoCAD VBA and Rhino Script, the lecture attempts to enhance the students' capability of reasonongly analyzing and solving design problems other than the skills of "macro" or "script" programming, to let them lay the base of digital architectural design.
The course consists of three parts:
1. Introduction to VB, including the basic grammar of VB, structural program, array, process, etc.
2. AutoCAD VBA, including the structure of AutoCAD VBA , 2D graphics, interactive methods, 3D objects, and some basic knowledge of computer graphics.
3. Brief introduction of Rhino Script, including basic concepts, the concept of Nurbs, sammary of VBScript, and Rhino objects.

研究生一年级
GIS基础与应用・童滋雨
课程类型：选修
学时学分：18学时/1学分

Graduate Program 1st Year
CONCEPT AND APPLICATION OF GIS • TONG Ziyu
Type: Elective Course
Study Period and Credits:18 hours / 1 credits

课程介绍
本课程的主要目的是让学生理解GIS的相关概念以及GIS对城市研究的意义，并能够利用GIS软件对城市进行分析和研究。

Course Description
This course aims to enable students to understand the related concepts of GIS and the significance of GIS to urban research, and to be able to use GIS software to carry out urban analysis and research.

其他
MISCELLANEA

讲座

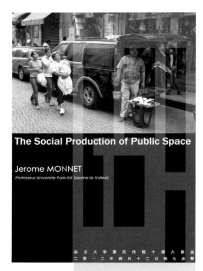

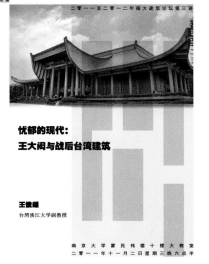

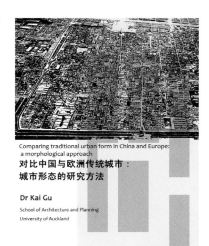

硕士学位论文列表

研究生姓名	研究生论文标题	导师姓名
陈鑫	福建土堡为代表的传统版筑建造体系研究	赵辰
吴汉成	皱褶形态模型及其在建筑设计中的应用研究	张雷
张东光	数字建筑及四个经典算法的分析与模拟	吉国华
谢明潭	十九世纪德国的"风格"论战	王骏阳
白杰	石材建构方式中的叠砌研究	张雷
陈晨	建筑创作中U型玻璃的应用表达及其局限性探究	冯金龙
陈秋菊	扬州广福花园拆迁安置小区设计及其Holcim全球可持续建筑大奖赛申报材料研究	张雷
陈卓然	建筑与景观整合策略的参数化实现初探	冯金龙
成敏	基于天空开阔度的城市街道空间形态研究——以南京为例	丁沃沃
段梦媛	控制性详细规划指标体系对居住街廓形态影响量化分析——以南京为例	吉国华
高林	二维表面形式研究及其在建筑中的应用	张雷
顾志勤	南京小户型公寓的户型设计研究	冯金龙
扈小璇	非均质肌理平均天空开阔度的研究——以南京为例	丁沃沃
黎思琪	实体面材可丽耐的设计与建造研究	傅筱
刘建岗	阿道夫·路斯独栋住宅空间演化的呈现及对穆勒住宅的解读	王骏阳
刘玮	"新建筑"改造研究——以建邺区体育大厦和南林大汽配城为例	胡恒
刘振强	泰州"勾连搭"做法研究	萧红颜
卢伟	2012年"中国大学生建造节"——南大建造教学的新探索	赵辰
缪纯乐	Revit自适应构件功能在形体与表皮建模方法上的研究	傅筱
阮玲	紧凑城市观念下香港公屋的建筑形态与规划结构研究	周凌

研究生姓名	研究生论文标题	导师姓名
沈晶晶	清末民初南京传统"楼宅"类型及相关技术初探	萧红颜
孙慧玲	城市更新背景下历史城区居住性街巷研究——以南京为例	赵辰
田野	建筑体形系数与街廓形态的关系初探	吉国华
汪杉	GRC板材在建筑设计中的应用研究	傅筱
王涵	基于DIVA模拟计算的外遮阳系数研究	吉国华
王星	基于分形理论的V形街道天际轮廓线模拟生成研究	吉国华
吴子夜	玻璃的材质加工和视觉表现力研究	张雷
夏芸	柴墟水景街区中三种建筑类型的设计研究	周凌
谢方洁	基于图解静力学的各建筑结构类型图析建立尝试	萧红颜
徐路华	国民革命军阵亡将士纪念塔的形制与技术探析	萧红颜
杨扬	城市粗糙度理论下南京居住小区风环境CFD模拟方法与肌理形态关系研究	丁沃沃
应超	当代大型公共图书馆共享空间的演变	周凌
章玲玲	另类诠释——近代域外建筑者在华之中国民间风格建筑创作研究	赵辰
张伟伟	ECOTECT与DesignBuilder在能耗模拟方面的比较研究	吴蔚
张熙慧	南湖新村三十年史	胡恒
张赟	隈研吾的建筑作品及理论研究	冯金龙
赵涵	基于风速的公共建筑围合的街道轮廓形态研究	丁沃沃
周文婷	城市更新视野中的苏州城市街坊外部空间风环境研究	华晓宁
祝贺	工业遗产保护再利用与城市空间一体化——以南京为例	赵辰

本科生 Undergraduate

2008级学生 Students 2008

陈 成 CHEN Cheng	黄文华 HUANG Wenhua	龙俊荣 LONG Junrong	陶敏悦 TAO Minyue	岳文博 YUE Wenbo
陈 丹 CHEN Dan	蒋菁菁 JIANG Jingjing	陆敏艳 LU Minyan	王冬雪 WANG Dongxue	张 成 ZHANG Cheng
陈 娟 CHEN Juan	姜小颖 JIANG Xiaoying	陆 恬 LU Tian	王洁琼 WANG Jieqiong	张方籍 ZHANG Fangji
陈牧野 CHEN Muye	鞠昌荣 JU Changrong	潘 东 PAN Dong	徐怡雯 XU Yiwen	张 伟 ZHANG Wei
陈 鹏 CHEN Peng	李可歆 LI Kexin	邱洋凯 QIU Yangkai	薛晓旸 XUE Xiaoyang	赵 芹 ZHAO Qin
冯琦勋 FENG Qixun	李文聪 LI Wencong	任承美 REN Chengmei	颜骁程 YAN Xiaocheng	郑国活 ZHENG Guohuo
耿 健 GENG Jian	林晓睿 LIN Xiaorui	邵一丹 SHAO Yidan	殷 强 YIN Qiang	朱龙腾 ZHU Longteng
胡 笳 HU Jia	林中格 LIN Zhongge	司秉丹 SI Binghui		

2009级学生 Students 2009

陈观兴 CHEN Guanxing	韩 霄 HAN Xiao	刘玉婧 LIU Yujing	王天宇 WANG Tianyu	杨 谦 YANG Qian
陈 凛 CHEN Lin	何家斌 HE Jiabin	浦雪飞 PU Xuefei	魏江洋 WEI Jiangyang	伊若琛 YI Ruochen
陈思逸 CHEN Siyi	吉永峰 JI Yongfeng	孙冠成 SUN Guancheng	吴 宾 WU Bin	乐 磊 YUE Lei
戴 波 DAI Bo	李 萍 JI Ping	谭发兵 TAN Fabing	夏冬冬 XIA Dongdong	郑金海 ZHENG Jinhai
戴 璟 DAI Jing	贾江南 JIA Jiangnan	汤建华 TANG Jianhua	徐 蕾 XU Lei	周 倩 ZHOU Qian
段艳文 DUAN Yanwen	蒯冰清 KUAI Bingqing	王适远 WANG Shiyuan	徐沁心 XU Qinxin	周荣楼 ZHOU Ronglou
顾三省 GU Sansheng	力振球 LI Zhenqiu			

2010级学生 Students 2010

陈博宇 CHEN Boyu	胡任元 HU Renyuan	刘树豪 LIU Shuhao	谭 健 TAN Jian	杨天仪 YANG Tianyi
陈凌杰 CHEN Lingjie	黄广伟 HUANG Guangwei	刘思彤 LIU Sitong	王 琳 WANG Lin	杨玉菡 YANG Yuhan
陈晓敏 CHEN Xiaomin	蒋 婷 JIANG Ting	刘文沛 LIU Wenpei	夏候蓉 XIA Hourong	姚 梦 YAO Meng
陈修远 CHEN Xiuyuan	李乐之 LI Lezhi	刘 宇 LIU Yu	徐 晏 XU Yan	张明杰 ZHANG Mingjie
程 斌 CHENG Bin	李平乐 LI Pingle	鲁光耀 LU Guangyao	许梦逸 XU Mengyi	周平浪 ZHOU Pinglang
顾一蝶 GU Yidie	林 治 LIN Zhi			

2011级学生 Students 2011

崔傲寒 CUI Aohan	蒋建昕 JIANG Jianxin	彭丹丹 PENG Dandan	王新宇 WANG Xinyu	张豪杰 ZHANG Haojie
冯琪 FENG Qi	蒋造时 JIANG Zaoshi	宋富敏 SONG Fumin	吴佳禾 WU Jiahe	张黎萌 ZHANG Limeng
顾聿笙 GU Yusheng	雷朝荣 LEI Zhaorong	拓展 TUO Zhan	吴结松 WU Jiesong	张人祝 ZHANG Renzhu
黄凯峰 HUANG Kaifeng	黎乐源 LI Leyuan	王梦琴 WANG Mengqin	席弘 XI Hong	周松 ZHOU Song
黄雯倩 HUANG Wenqian	柳纬宇 LIU Weiyu	王却骏 WANG Quelian	谢忠雄 XIE Zhongxiong	周贤春 ZHOU Xianchun
贾福龙 JIA Fulong	缪姣姣 MIAO Jiaojiao	王思绮 WANG Siqi	徐亦杨 XU Yiyang	左思 ZUO Si
蒋佳瑶 JIANG Jiayao	倪若宁 NI Ruoning	王潇聆 WANG Xiaoling	杨益晖 YANG Yihui	

研究生 Postgraduate

白 杰 BAI Jie	陈卓然 CHEN Zhuoran	扈小璇 HU Xiaoxuan	卢 伟 LU Wei	田 野 TIAN Ye	夏 芸 XIA Yun	应 超 YING Chao	张 奘 ZHANG Yun	
曹庆艳 CAO Qingyan	成 敏 CHENG Min	黎思琪 LI Siqi	缪纯乐 MIAO Chunle	汪 杉 WANG Shan	谢方洁 XIE Fangjie	张厚亮 ZHANG Houliang	赵 涵 ZHAO Han	
陈 晨 CHEN Chen	段梦媛 DUAN Mengyuan	刘建岗 LIU Jiangang	阮 玲 RUAN Ling	王 涵 WANG Han	谢明潭 XIE Mingtan	章玲玲 ZHANG Lingling	周文婷 ZHOU Wenting	
陈 娟 CHEN Juan	高 林 GAO Lin	刘 玮 LIU Wei	沈晶晶 SHEN Jingjing	王 星 WANG Xing	徐路华 XU Luhua	张伟伟 ZHANG Weiwei	祝 贺 ZHU He	
陈秋菊 CHEN Qiuju	顾志勤 GU Zhiqin	刘振强 LIU Zhenqiang	孙慧玲 SUN Huiling	吴子夜 WU Ziye	杨 扬 YANG Yang	张熙慧 ZHANG Xihui		

鲍丽丽 BAO Lili	李日影 LI Riying	罗思维 LUO Siwei	闻金石 WEN Jinshi	朱俊杰 ZHU Junjie	洪海燕 HONG Haiyan	李 倩 LI Qian	陆 磊 LU Lei	杨 萍 YANG Ping
谌 利 CHEN Li	李善超 LI Shanchao	金筱敏 JIN Xiaomin	吴 宁 WU Ning	朱 珠 ZHU Zhu	黄佳秋 HUANG Jiaqiu	李亚楠 LI Ya'nan	潘 慧 PAN Hui	叶 鹏 YE Peng
程 路 CHENG Lu	李莹莹 LI Yingying	汤梦捷 TANG Mengjie	吴仕佳 WU Shijia	鲍 强 BAO Qiang	黄 婧 HUANG Jing	李艳丽 LI Yanli	曲 亮 QU Liang	易 微 YI Wei
丁文博 DING Wenbo	李苑常 LI Yuanchang	涂梦如 TU Mengru	吴 玺 WU Xi	鲍颖峰 BAO Yingfeng	江 萌 JIANG Meng	刘杜娟 LIU Dujuan	王 健 WANG Jian	张万金 ZHANG Wanjin
丁文磊 DING Wenlei	林天予 LIN Tianyu	王鑫星 WANG Xinxing	俞 英 YU Ying	陈文龙 CHEN Wenlong	蒋 敏 JIANG Min	刘 杰 LIU Jie	王素玉 WANG Suyu	赵金林 ZHAO Jinlin
郭 芳 GUO Fang	刘 昀 LIU Jun	王雅谦 WANG Yaqian	曾宇城 ZENG Yucheng	陈雯茜 CHEN Wenqian	李广林 LI Guanglin	刘亚楠 LIU Ya'nan	伍蔡畅 WU Caichang	祝 凯 ZHU Kai
黄志鹏 HUANG Zhipeng	柳 楠 LIU Nan	王 一 WANG Yi	赵 慧 ZHAO Hui	董振宇 DONG Zhenyu	李 卉 LI Hui			

陈 肯 CHEN Ken	柯国新 KE Guoxin	马 喆 MA Zhe	吴绉彦 WU Zhouyan	袁 芳 YUAN Fang	葛鹏飞 GE Pengfei	倪力均 NI Lijun	王 晨 WANG Chen	张 岸 ZHANG An
陈 圆 CHEN Yuan	黎健波 LI Jianbo	孟庆忠 MENG Qingzhong	谢智峰 XIE Zhifeng	张 备 ZHANG Bei	胡 曜 HU Yao	潘 旻 PAN Min	夏 澍 XIA Shu	张 敏 ZHANG Min
陈 钊 CHEN Zhao	李 港 LI Gang	沈周娅 SHEN Zhouya	辛胤庆 XIN Yinqing	张卜予 ZHANG Buyu	金 鑫 JIN Xin	彭文楷 PENG Wenkai	徐庆姝 XU Qingshu	张 培 ZHANG Pei
高 菲 GAO Fei	李恒鑫 LI Hengxin	石延安 SHI Yan'an	徐 睿 XU Rui	曹梦原 CAO Mengyuan	李红瑞 LI Hongrui	乔 力 QIAO Li	姚丛琦 YAO Congqi	张永雷 ZHANG Yonglei
管 理 GUAN Li	刘滨洋 LIU Binyang	王海芹 WANG Haiqin	杨尚宜 YANG Shangyi	陈 姝 CHEN Shu	李 扬 LI Yang	邱金宏 QIU Jinhong	于海平 YU Haiping	赵 锐 ZHAO Rui
韩 梦 HAN Meng	刘兴渝 LIU Xingyu	王力凯 WANG Likai	殷 奕 YIN Yi	陈婷婷 CHEN Tingting	刘 宇 LIU Yu	沈均臣 SHEN Junchen	虞王璐 YU Wanglu	赵天亚 ZHAO Tianya
胡 昊 HU Hao	刘奕彪 LIU Yibiao	王亦播 WANG Yibo	郁新新 YU Xinxin	陈 新 CHEN Xin	吕 程 LÜ Cheng	汪 园 WANG Yuan	袁金燕 YUAN Jinyan	周 逸 ZHOU Yi
胡鹭鹭 HU Lulu	吕 铭 LÜ Ming							